The Transmission-Line Modeling (TLM) Method in Electromagnetics

The Transmission-Line Modeling (TLM) Method in Electromagnetics
Christos Christopoulos

978-3-031-00563-3   paperback   Christopoulos
978-3-031-01691-2   ebook   Christopoulos

DOI 10.1007/978-3-031-01691-2

A Publication in the Springer series
SYNTHESIS LECTURES ON COMPUTATIONAL ELECTROMAGNETICS
Lecture #7
Editor: Constantine A. Balanis, Arizona State University

Series ISSN: Print 1932-1252 Electronic 1932-1716

First Edition
10 9 8 7 6 5 4 3 2 1

The Transmission-Line Modeling (TLM) Method in Electromagnetics

Christos Christopoulos
Professor of Electrical Engineering and Director,
George Green Institute for Electromagnetics Research,
University of Nottingham, UK

SYNTHESIS LECTURES ON COMPUTATIONAL ELECTROMAGNETICS #7

ABSTRACT

This book presents the topic in electromagnetics known as Transmission-Line Modeling or Matrix method-TLM. While it is written for engineering students at graduate and advanced undergraduate levels, it is also highly suitable for specialists in computational electromagnetics working in industry, who wish to become familiar with the topic. The main method of implementation of TLM is via the time-domain differential equations, however, this can also be via the frequency-domain differential equations. The emphasis in this book is on the time-domain TLM. Physical concepts are emphasized here before embarking onto mathematical development in order to provide simple, straightforward suggestions for the development of models that can then be readily programmed for further computations. Sections with strong mathematical flavors have been included where there are clear methodological advantages forming the basis for developing practical modeling tools. The book can be read at different depths depending on the background of the reader, and can be consulted as and when the need arises.

KEYWORDS

Computational electromagnetics, Matrix method-TLM, Time-domain methods, Transmission-line modeling, TLM.

For Katherine, Alex, and Michael

Contents

Preface

This book is designed to present the topic of TLM to a graduate or advanced undergraduate audience. It will also be suitable for computational electromagnetics specialists working in industry who wish to become familiar with this technique. I have tried to emphasize physical concepts first before embarking into a mathematical development and to give simple straightforward suggestions for the development of models that can then be readily programed for further computation. I have, however, included sections with a strong mathematical flavor where I felt that there were clear methodological advantages and where they form the basis for developing practical modeling tools. Some of these sections could be omitted on first reading to allow students to develop simple programs in a short period of time without the effort necessary to understand some of the theoretical underpinnings of the subject. I refer in particular to Sections 4.5, 4.6, Chapter 5 and Sections 6.5–6.7.

I have given a number of references that complement the material in this text and can be used to broaden and deepen understanding. I hope that the text can be read at different depths depending on the background of the reader and can be revisited as and when the need arises.

I acknowledge here the contributions to my understanding of the subject made by numerous colleagues who have worked with me at Nottingham and in other laboratories over a number of years. Most of them are mentioned as authors in the references quoted. I am indebted to them for helping me to gain a better insight into TLM and for their friendship.

Christos Christopoulos
Nottingham, January 2006

CHAPTER 1

Modeling as an Intellectual Activity

1.1   AN INTRODUCTION TO MODELING

Humans have an innate ability to formulate "models" of the world around them so that they can classify incidents, predict likely outcomes, and respond intelligently to their environment. It can be argued that the success of humans in adapting to and changing their environment is due primarily to their modeling ability. At the highest level models are conceptual, that is, involve observations of regularities in nature and hence the designation of broad categories and classifications of similar events. In many cases this is all that is required to formulate intelligent responses to events. In such cases, although an explicit model has not been formulated as such, there is, however, a rudimentary model in operation, assisting in the interpretation of events and the formulation of responses. Human knowledge and wisdom are intricately associated with the formulation and selection of the appropriate model or models and the appropriate action in response.

In scientific and technical fields, where a detailed quantitative response is required, models are explicit and considerably more detailed to account for a multitude of inputs, and to formulate designs and develop detailed quantitative responses. They may be physical models, e.g., a mannequin to allow proper tailoring of a dress, or analytical, e.g., Ohm's law relating potential difference and current across a resistor and may reside in algorithms embedded in computer software. All such models share the following characteristics:

- Models are not the real things and neither should they be! They should be good enough for the class of phenomena to be studied and no more. If they are more complex and powerful than they need to be, then they will lack clarity and will not be as useful and versatile as desired. An ideal model is one that is so simple as to be capable of use while travelling home after work, on the train, scribbling at the back of an envelope. Alas, such models are very difficult to formulate in the complex world we live in. But it must be stressed that we need models that have clarity, physical transparency, are easy to use and hence assist creative thought.

- In most cases the above objectives can be best met by having several models for the same physical entity. As an illustration I mention examples of several models of a human, e.g., a mannequin for fitting clothes, a collection of masses and springs for studying vibration of the human frame during flight, a 100 pF capacitor in series with a 1.5 kΩ resistor for studying electrostatic charging, etc. Each one of these models is very good for the job it is intended to do but they fail badly if applied outside their domains of validity. A crucial skill of any user of models is to be able to select the correct one and know when the model ceases to be valid—no model is valid under all circumstances!

Our task in these lectures is to develop models for a particular class of phenomena but we must, from the very start, be aware that whatever we do has limitations and we must learn how to cope with these limitations. Otherwise, a model at the hands of the uninitiated, however "user friendly" it is and whatever pretty graphics it produces is a deeply flawed and even dangerous tool.

Let me set out in broad terms the context in which every modeling activity takes place [1]:

- Confronted with a physical event one needs to establish the broad domain of science applicable to this event. This process is known as *conceptualization* and in our case it is likely to lead us to the concept of electromagnetic interactions.

- Following the identification of the relevant concepts a more mathematical *formulation* is required (e.g., Maxwell's equations).

- The mathematical model thus formulated is then prepared for solution by computer by designing an appropriate *numerical implementation*.

- The development of software for the final *computation* of the problem is then based on these numerical algorithms.

- Finally, the results of the computation must be checked for physical reasonableness, compared with other similar solutions or experiments if possible so that a process of *validation* is implemented.

If validation is not satisfactory, then a complete rethink of all the previous stages may be necessary. Clearly, modeling as seen through these distinct stages, is a lengthy and demanding process requiring a set of skills and intellectual attributes of the highest order. Only then can the results of simulations based on the model can be used with confidence to provide an aid to analysis and design of complex systems. It may be argued that many simulations do not have to go through all these detailed processes outlined above. This is true if one refers to an explicit implementation of each step. However, implicitly every modeller considers every step in the process but inevitably focuses on one or more specific aspects. What I am trying to stress is that simulation is not merely about writing computer programmes. It is the entire process outlined above and it is of value only if all elements are balanced to reflect the particular aspects of each problem.

1.2   TYPES OF MODELS USED IN ELECTROMAGNETICS

Most electromagnetic models can be conveniently classified into two broad categories in an effort to identify generic advantages and disadvantages of particular models. In formulating a problem in electromagnetics one often arrives at an expression relating a stimulus function $g(x)$ to a response function $\Phi(x)$. The two are related through an operator,

$$L\left\{\Phi(x)\right\} = g(x) \tag{1.1}$$

where L is the *operator* and x is the *variable*. Numerical methods may be conveniently classified according to the *domain* of the operator and the variable. We may have thus a variable which is either time (time-domain or TD method), or frequency (frequency-domain or FD method). Similarly, the operator may be a either differential one (differential equation or DE method) or an integral one (integral equation or IE method). Some very high frequency methods such as those based on ray techniques do not fit these classifications but the great majority of modeling techniques do.

We can thus have several generic models, e.g., a time-domain integral equation method (TD-IE), etc. The most popular combinations so far are TD-DE methods and FD-IE methods. Examples of the former are the Finite-Difference Time-Domain (FD-TD) method [2, 3], the Transmission-Line Modeling or Matrix (TLM) method [4] and that of the latter the Method of Moments (MoM) [5].

The choice of the domain of the operator and of the variable affect significantly the characteristics of each technique. A TD method is inherently suitable for studying transients, wide-band applications, and nonlinear problems. An FD is better suited to steady-state narrow-band applications. Naturally, through the use of Fourier Transforms one can go from frequency-domains to the time-domains but this is done at some cost. Sampling in space for a DE operator normally requires a full-volume discretization which is computationally expensive but offers scope for dealing in detail with nonuniform materials and complex geometrical features. In contrast, for IE operators discretization is normally required on important surfaces of the problem and thus computation may be cheaper. However, the flexibility of dealing with intricate geometrical detail and/or inhomogeneous materials is curtailed. The important conclusion stemming from these observations is that whatever generic formulation is employed it bequeaths inherited characteristics, some advantageous and some not, and no method excels in every conceivable practical situation. Recognition of this fact by users is essential for avoiding sterile arguments about "which method is best" and in selecting the best method for each particular task.

In the succeeding chapters I will describe a particular method known as the Transmission-Line Modeling or Matrix method-TLM in short. The word "Matrix" is meant to refer to the use of a matrix of transmission lines as the modeling medium but should not be confused with the use of matrices to solve a problem—no matrix inversion is required in TLM. For this reason I prefer the term "modeling" to "matrix". The main implementation of TLM is as a time-domain differential equation method but there are also TLM implementations in the frequency-domain. The emphasis of this book is on the time-domain TLM.

CHAPTER 2

Field and Network Paradigms

2.1 FROM FIELDS TO NETWORKS

I have discussed in Chapter 1 the human imperative of constructing models of our environment so that we can formulate rapid and effective responses to events. I also pointed out that a hierarchy of models is necessary, even for the same physical entity, each tuned to a particular task. No model is worse than another—it simply happens that sometimes users employ particular models outside their range of applicability. A model may be more sophisticated or complex than another one but this does not make it "better." Common sense dictates that we do not "use a sledge hammer to crack a nut." It is the same with models for electromagnetic phenomena, we need a hierarchy of models with a range of complexity to address particular problems. The objective of this chapter is to discuss what these models may be for the study of electrical phenomena.

Students of electrical engineering at University are taught extensively how to solve "circuits" using network theorems such as Kirchhoff's laws. Indeed, the first exposure to electrical work is through the development of these theorems. It is only later that field concepts are introduced. This sequence of events is dictated by convenience as network concepts are simpler to absorb and the mathematical skills required are far less sophisticated compared to those required for field work. A lot of worthwhile problems can be tackled using network concepts, with modest difficulty, offering a straightforward and rapid introduction to electrical phenomena. This convenience, however, obscures the fundamental nature of field concepts and makes them appear to many as an unnecessary complexity brought upon the otherwise simple and ordered world of the network-based view of electrical phenomena. Some, happily very few, even believe that the introduction of field concepts is perpetrated upon students for the sole purpose of offering sadistic pleasures to professors! It is therefore of some significance to outline clearly under what circumstances network or alternatively field concepts prevail. Since TLM in a sense is a bridge between the two, this discussion will be particularly useful as model building in TLM proceeds in the following chapters.

At low frequencies, circuits are understood in terms of lumped components (e.g., capacitors, inductors, resistors) and quantities such as voltage and current. Low frequency here should not be understood in an absolute sense. What matters is the electrical size of the circuit, i.e., its dimensions relative to the wavelength λ at the highest frequency of interest. If the physical dimensions of the circuit in all directions are much smaller than the wavelength then the circuit operates in the low-frequency regime. In this regime, the appropriate paradigm is that of the "network" where the important factor in circuit operation is its topology [6]. The fundamental laws that constitute the mathematical framework of the network model are Kirchoff's laws.

If the circuit is electrically large (size comparable to the wavelength) only in one dimension, then we can talk about a "distributed" circuit and use a development of the network paradigm known as "transmission-line theory." This allows, under certain circumstances, propagation of waves guided by wires to be studied using essentially network-based techniques.

When the combinations of frequency and physical size are such that the circuit dimensions are comparable or larger than the wavelength in all three dimensions then the network paradigm fails. A network-based model used in this regime will result in uncontrollable errors and gross misconceptions. In this regime, the "field" paradigm is necessary. In employing the field concept we assign quantities to space so that the field transfers action between electrically charged matter. The mathematical embodiment of field concepts is in Maxwell's equations which are the starting point for models based on the field paradigm. The operation of a circuit at high frequencies where field concepts prevail depends not only on its topology (i.e., connectivity) but also on its geometry. This makes field problems considerably more difficult to solve than network problems.

We see from this discussion that the fundamental concept is that of the field. Network laws such as Kirchoff's laws are derived from Maxwell's equation under certain simplifying assumptions. Whenever possible we stay with network concepts, since they are much simpler to implement in models, but for electrically large structures we must recognize the need for new models where the special requirements of field behavior must be fully accounted for.

2.2   NETWORK ANALOGS OF PHYSICAL SYSTEMS

It is undeniable that a network model is simple, familiar to most, and efficient in its use. It would therefore appear that networks are the ideal models for those requiring a tool to test and develop their creative and design ideas.

There is no doubt that confronted with new situation humans seek to interpret it by reducing it to a familiar one. It was natural therefore that at the beginning of electrical science explanatory models of electrical phenomena were constructed based on mechanical models. This led to the conceptual development of lumped components such as L and C to describe concentrated energy storage in magnetic and electric fields, something akin to the storage of kinetic and potential energy in mechanical systems. It is therefore not unnatural that lumped components could be used to model other physical situations (including mechanical systems!) provided that appropriate care is taken. We are particularly interested in the use of lumped components to model fields. This at first may appear paradoxical in view of what we have said before, but let us try.

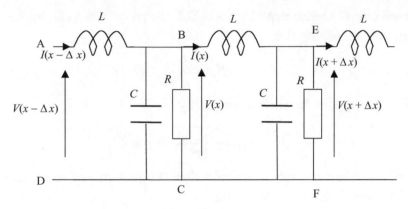

FIGURE 2.1: A cascade of segments part of a transmission line

We consider a long distributed circuit (a transmission line, TL) as shown in Fig. 2.1. Following normal practice, we "model" the distributed circuit as a cascade of segments each consisting of lumped components R, L, C representing dissipation and energy storage in magnetic and electric fields. We postpone a discussion of the length Δx of each segment to the next section. Applying Kirchoff's voltage (KVL) and current (KCL) laws on a segment and assuming $\Delta x \to 0$ we obtain

$$-\frac{\partial v(x, t)}{\partial x}\Delta x = L\frac{\partial i(x, t)}{\partial t} \tag{2.1}$$

$$-\frac{\partial i(x, t)}{\partial x}\Delta x = C\frac{\partial v(x, t)}{\partial t} + \frac{v(x, t)}{R} \tag{2.2}$$

Combining these two equations to eliminate the voltage we obtain

$$\frac{\partial^2 i(x, t)}{\partial x^2} = \frac{LC}{(\Delta x)^2}\frac{\partial^2 i(x, t)}{\partial t^2} + \frac{L}{(\Delta x)^2 R}\frac{\partial i(x, t)}{\partial t} \tag{2.3}$$

We will now contrast this with the propagation of a plane wave along the x-direction in a lossy medium such that

$$\boldsymbol{E} = (0, E_y, 0) \qquad \boldsymbol{H} = (0, 0, H_z)$$

For convenience Maxwell's equations are given in Appendix 1. For this case Faraday's law reduces to

$$\frac{\partial E_y(x, t)}{\partial x} = -\mu\frac{\partial H_z(x, t)}{\partial t} \tag{2.4}$$

and Ampere's law to

$$-\frac{\partial H_z(x, t)}{\partial x} = j_y + \varepsilon\frac{\partial E(x, t)}{\partial t} \tag{2.5}$$

Differentiating (2.4) with respect to x and (2.5) with respect to t, eliminating the magnetic field intensity and recognizing that

$$E_y = \frac{j_y}{\sigma}$$

gives the following equation for the current density

$$\frac{\partial^2 j_y(x, t)}{\partial x^2} = \mu\varepsilon\frac{\partial^2 j_y(x, t)}{\partial t^2} + \mu\sigma\frac{\partial j_y(x, t)}{\partial t} \tag{2.6}$$

where μ, ε, and σ are the magnetic susceptibility, dielectric permittivity, and electric conductivity of the material in which the wave propagates.

The key observation here is that Eqs. (2.3) for the cascade of networks and (2.6) for one-dimensional wave propagation in a lossy medium have exactly the same form. Waves in (2.6) propagate with a velocity

$$u = \frac{1}{\sqrt{\mu\varepsilon}} \tag{2.7}$$

Similarly, (2.3) indicates propagation with a velocity

$$u = \frac{1}{\sqrt{L/\Delta x \, C/\Delta x}} \tag{2.8}$$

This isomorphism offers a clue as to how the network in Fig. 2.1 may be the basis of a model for the field problem described by Maxwell's equations leading to (2.6). It is clear that if one can find a way to solve (2.3) the solution of (2.6) is readily obtained by invoking the analogies shown in Fig. 2.2.

The analogy is physically transparent as the inductance per unit length represents the magnetic permeability, the capacitance the dielectric permittivity, and the resistance the electric conductivity. One can establish an analogy between voltages and currents in the network with electric and magnetic fields. In subsequent chapters we will see how the one-dimensional analog in Fig. 2.1 can be extended to two and three dimensions. What we have done so far is to demonstrate how a field in continuous space can be "mapped" onto a network consisting of a cascade of sections each Δx long. We have effectively discretized the problem in *space*. It remains

Network	Field
i	j_y
$L/\Delta x$	μ
$C/\Delta x$	ε
$1/(R\Delta x)$	σ

FIGURE 2.2: Analogy between network and field quantities

to show how to discretize in *time*. We need to do both if we have to obtain a numerical solution. Otherwise, without a finite spatial length we will require an infinite number of memory storage locations and similarly without a finite sampling time Δt we would need infinitely long run times! Therefore, inherent to the numerical solution is the choice of the spatial Δx and temporal Δt sample lengths. In this section spatial sampling is achieved by lumping together capacitive and inductive properties. Sampling in time will be described in the next chapter. Before we do this, however, we need to examine a bit more the implications of spatial sampling. This we do in the next section.

2.3   THE IMPACT OF SPATIAL SAMPLING

In deriving Eqs. (2.1–2.2) we have assumed that $\Delta x \to 0$. However, we know this cannot be in a numerical solution. We can make the spatial sampling small but never zero. What are the implications of a finite spatial sampling length? In order to assess its impact we will apply again more carefully Kirchoff's laws in the network shown in Fig. 2.1 [7]. In order to illustrate more clearly the impact of discretization we will neglect losses ($R \to \infty$).

We apply KVL to loops ABCDA and BEFCB to obtain

$$v(x - \Delta x, t) = L\frac{\partial i(x - \Delta x, t)}{\partial t} + v(x, t) \qquad (2.9)$$

$$v(x, t) = L\frac{\partial i(x, t)}{\partial t} + v(x + \Delta x, t) \qquad (2.10)$$

Subtracting and rearranging (2.9) and (2.10) we obtain

$$2v(x, t) - v(x + \Delta x, t) - v(x - \Delta x, t) = L\frac{\partial}{\partial t}\left[i(x, t) - i(x - \Delta t, t)\right] \qquad (2.11)$$

We now apply KCL at node B (neglecting current in R) to obtain

$$i(x - \Delta x, t) = i(x, t) + C\frac{\partial v(x, t)}{\partial t} \qquad (2.12)$$

We now substitute (2.12) into (2.11) to eliminate the currents

$$LC\frac{\partial^2 v(x, t)}{\partial t^2} = v(x + \Delta x, t) + v(x - \Delta x, t) - 2v(x, t) \qquad (2.13)$$

Equation (2.13) is the equivalent to (2.3) where we have neglected losses and have not made the simplification of the space discretization tending to zero. We thus ended up with a difference equation rather than a differential equation. Let us now assume a wavelike dependence for the voltage and explore further (2.13). We start with

$$v(x, t) = V_{\text{pk}} \sin(\omega t - kx) \qquad (2.14)$$

where ω is the angular frequency and $k = 2\pi/\lambda$. This expression represents waves propagating with a phase velocity

$$u = \frac{\omega}{k} \tag{2.15}$$

We can establish how ω and k are related (and therefore how u varies with frequency) by substituting (2.14) into (2.13). We will obtain what is known as the *dispersion relation* for waves propagating in the network of Fig. 2.1.

By differentiating $v(x, t)$ in (2.14) twice we obtain

$$\frac{\partial^2 v(x, t)}{\partial t^2} = -\omega^2 V_{\text{pk}} \sin(\omega t - kx) \tag{2.16}$$

We now proceed with the evaluation of terms on the RHS of (2.13)

$$
\begin{aligned}
v(x + \Delta x, t) - v(x, t) &= V_{\text{pk}} \left\{ \sin[(\omega t - kx) - k\Delta x] - \sin[\omega t - kx] \right\} \\
&= V_{\text{pk}} \left\{ \sin(\omega t - kx)\cos(k\Delta x) - \cos(\omega t - kx)\sin(k\Delta x) - \sin(\omega t - kx) \right\} \\
&= V_{\text{pk}} \left\{ \sin(\omega t - kx)[\cos(k\Delta x) - 1] - \cos(\omega t - kx)\sin(k\Delta x) \right\} \\
&= V_{\text{pk}} \left\{ \sin(\omega t - kx)\left[-2\sin^2\left(\frac{k\Delta x}{2}\right)\right] - \cos(\omega t - kx)2\sin\left(\frac{k\Delta x}{2}\right)\cos\left(\frac{k\Delta x}{2}\right) \right\} \\
&= -2V_{\text{pk}} \sin\left(\frac{k\Delta x}{2}\right)\cos\left[\omega t - \left(kx + \frac{k\Delta x}{2}\right)\right]
\end{aligned}
\tag{2.17}
$$

In a similar fashion, we evaluate

$$v(x - \Delta x, t) - v(x, t) = 2V_{\text{pk}} \sin\left(\frac{k\Delta x}{2}\right)\cos\left[\omega t - \left(kx - \frac{k\Delta x}{2}\right)\right] \tag{2.18}$$

Therefore, by adding (2.17) and (2.18) the RHS of (2.13) becomes

$$v(x + \Delta x, t) + v(x - \Delta x, t) - 2v(x, t) = -4V_{\text{pk}} \sin^2\left(\frac{k\Delta x}{2}\right)\sin(\omega t - kx) \tag{2.19}$$

Substituting (2.16) and (2.19) into (2.13) we obtain

$$\omega^2 = \frac{1}{LC}4\sin^2\left(\frac{k\Delta x}{2}\right) \tag{2.20}$$

Equation (2.20) is the required dispersion relation and it is clear that interdependence of ω and k is complicated and is the one in which the discretization length plays a role. On a continuous transmission line ($\Delta x \to 0$) all frequencies propagate at the same velocity given by (2.8).

However, for finite spatial sampling ($\Delta x \neq 0$) the velocity of propagation is frequency dependent as implied by (2.20). A square pulse launched on a line with the dispersion relation of (2.20) will disperse, as its various constituent frequency components will travel at different speeds. The process of discretization (spatial sampling) has therefore resulted in errors (numerical dispersion). Naturally, some lines exhibit inherent dispersion but the dispersion discussed here is entirely numerical in nature and therefore undesirable. It remains to explore further the magnitude of dispersion errors so that we get an understanding of how small Δx needs to be for numerical dispersion errors to be kept small. If we can assume that

$$\frac{k\Delta x}{2} \ll 1 \qquad (2.21)$$

We can approximate the sin by its argument

$$\sin\left(\frac{k\Delta x}{2}\right) \simeq \frac{k\Delta x}{2} \qquad (2.22)$$

and (2.20) reduces to

$$\left(\frac{\omega}{k}\right)^2 = \frac{(\Delta x)^2}{LC} \qquad (2.23)$$

Equation (2.23) is identical to (2.8) for the continuous case and therefore the numerical dispersion error disappears. Since,

$$k\Delta x = \frac{2\pi}{\lambda}\Delta x$$

Condition (2.21) is tantamount to saying that the sampling length must be much smaller than the wavelength at the highest frequency of interest. Only then, can we assume that numerical dispersion is negligible and therefore that the discrete network is an acceptable representation of the actual system. Errors are always present but by using fine enough discretization we can minimize their impact.

The dispersion analysis presented here for one-dimensional propagation is complicated enough but it can get almost intractable for more complex three-dimensional networks. Establishing the dispersion properties of a numerical scheme can be very difficult. We have given here a simple example to illustrate the impact of spatial sampling and the main conclusion that holds irrespective of the dimensionality of the network is that the sampling length must be smaller than the wavelength at the highest frequency of interest. A useful "rule of thumb" is that

$$\Delta x \le \frac{\lambda}{10} \qquad (2.24)$$

In the next section we will explore how sampling in time is accomplished and what its impact is.

CHAPTER 3

Transmission Lines and Transmission-Line Models

3.1 TRANSMISSION LINES

In the last chapter we have introduced the idea of networks consisting of lumped components as a way of discretizing electrical phenomena in space. It is now time to investigate how discretization in time may be accomplished. Central to this in TLM is the transmission-line (TL) theory. Therefore, we summarize here some of the essential TL results [7, 8].

We focus on lossless lines as they illustrate the basic concepts. Losses may be introduced at any time either in series or shunt configuration with the effect of making the propagation constant and the characteristic impedance complex quantities.

A lossless TL segment needs, in addition to its length ℓ, two quantities for its full characterization. These most fundamentally are its inductance L_d and capacitance C_d per unit length. Alternatively, this fundamental pair may be substituted by the propagation velocity u and characteristic impedance Z of the line. Yet another combination is the transit time along the line τ and its characteristic impedance Z. These parameters are not independent. For the TL segment in Fig. 3.1(a) of length ℓ and the per unit length inductance and capacitance shown, a potential difference V impressed at one end will propagate say a distance Δx in time Δt, hence the charge transferred to charge this section of the TL is

$$\Delta Q = (C_d \Delta x)V \qquad (3.1)$$

The current will therefore be

$$I = \frac{\Delta Q}{\Delta t} = C_d V \frac{\Delta x}{\Delta t} = C_d V u \qquad (3.2)$$

Similarly, the magnetic flux linked with this fraction of the TL inductance, which is charged, is

$$\Delta \Phi = (L_d \Delta x)I = L_d \Delta x C_d V u$$

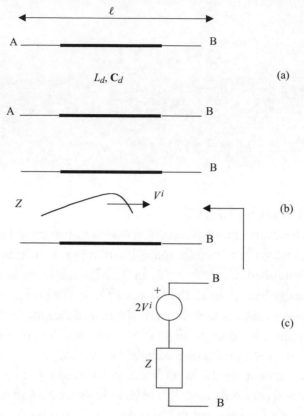

FIGURE 3.1: Transmission-line segment (a), with a voltage pulse V^i travelling towards BB (b), and Thevenin equivalent seen at BB (c)

where (3.2) is used to substitute for the current. From Faraday's law we obtain

$$V = \frac{\Delta \Phi}{\Delta t} = L_d C_d V u^2 \tag{3.3}$$

and eliminating V we obtain an expression for the velocity of propagation

$$u = \frac{1}{\sqrt{L_d C_d}} \tag{3.4}$$

Substituting (3.4) into (3.2) gives the relationship between the voltage and current pulses on the line

$$I = \frac{V}{\sqrt{L_d/C_d}} = \frac{V}{Z} \tag{3.5}$$

We note that the velocity of propagation on the TL is entirely dependent on the material properties, i.e.,

$$u = \frac{1}{\sqrt{L_d C_d}} = \frac{1}{\sqrt{\mu \varepsilon}} \qquad (3.6)$$

However, the characteristic impedance of the TL is not simply equal to the intrinsic impedance of the medium, i.e.,

$$Z = \sqrt{\frac{L_d}{C_d}} \neq \eta = \sqrt{\frac{\mu}{\varepsilon}} \qquad (3.7)$$

Z in addition to the material properties depends on the dimensions of the line. The transit time of the line is

$$\tau = \ell/u = \ell\sqrt{L_d C_d} = \sqrt{LC} \qquad (3.8)$$

where L and C are the total inductance and capacitance of the TL segment. The characteristic impedance is similarly

$$Z = \sqrt{\frac{L/\ell}{C/\ell}} = \sqrt{\frac{L}{C}} \qquad (3.9)$$

Multiplying and dividing (3.8) and (3.9) we obtain

$$\tau Z = L, \qquad \tau/Z = C \qquad (3.10)$$

We see that setting the transit time and the characteristic impedance of a TL segment defines a capacitance and inductance given by (3.10).

It can be shown that during the charging of the TL half the energy supplied by the source is stored in the electric field (capacitance) and half in the magnetic field (inductance). If we assume that the far end of the segment is an open circuit then the boundary condition there dictates that the current must be brought to zero. This is the origin of the reflection at the open circuit which takes the form of voltage and current waves travelling towards the source end such that the total current on the line is brought to zero. Therefore, since the line is assumed lossless, the energy stored in the inductance can no longer stay there ($I = 0$) and therefore the voltage doubles to accommodate in the capacitance the half of the total stored energy that has been associated with the inductance. In summary, a voltage pulse travelling on a TL and impinging on an open circuit doubles in size! This is a useful result as it allows us to construct the Thevenin equivalent circuit of a TL segment which is valid for a time interval equal to τ. The situation is illustrated in Fig. 3.1(b) showing a voltage pulse V^i incident on the port BB. We assume that the characteristic impedance of the TL is Z and that a new pulse may be injected into the line at port AA at regular intervals τ. The implication of this is that during each interval τ the pulse travelling on the TL does not change and therefore it is legitimate to assert that looking into the TL at BB the open-circuit voltage required by the Thevenin equivalent circuit is $2V^i$. Similarly, over the time interval τ a voltage V injected into the line at BB will draw a current equal to V/Z since at the injection point there can be no knowledge of the actual TL termination at AA. Therefore, the input impedance of the line during the time interval τ is Z. The complete equivalent circuit is shown in Fig. 3.1(c). We note that this treatment refers to the pulsed excitation of the line. If the line is subject to a harmonic excitation and we are interested in its steady-state response then the voltage and current phasors at the start and at the end of the TL segment are related by the expression

$$
\begin{bmatrix} \bar{V}(0) \\ \bar{I}(0) \end{bmatrix} = \begin{bmatrix} \cos(\beta\ell) & jZ\sin(\beta\ell) \\ j\dfrac{\sin(\beta\ell)}{Z} & \cos(\beta\ell) \end{bmatrix} \begin{bmatrix} \bar{V}(\ell) \\ \bar{I}(\ell) \end{bmatrix} = \begin{bmatrix} A & B \\ C & D \end{bmatrix} \begin{bmatrix} \bar{V}(\ell) \\ \bar{I}(\ell) \end{bmatrix}
\tag{3.11}
$$

where $\beta = \omega\sqrt{L_d C_d}$ is the phase constant of the line and therefore $\beta\ell = \omega\tau$. The matrix relating input and output parameters is known as the ABCD matrix. The input impedance Z_{in} of the TL under these conditions is a complicated function given by the formula

$$
\frac{Z_{\text{in}}}{Z} = \frac{\dfrac{Z_1}{Z} + j\tan(\beta\ell)}{1 + j\dfrac{Z_1}{Z}\tan(\beta\ell)}
\tag{3.12}
$$

where Z_1 is the impedance of the load connected at the end of the line.

3.2   TIME DISCRETIZATION OF A LUMPED COMPONENT MODEL

We illustrate as an example the way in which conditions on a capacitor can be converted into time discrete form. In broad terms we reason that a transmission-line segment consists of distributed inductance and capacitance and therefore it may be possible to emphasize its capacitive aspects, in order to model a capacitor C, while minimizing its inductive nature. It is also clear that however we accomplish this, there will still remain some inductive contribution that must be interpreted as a modeling error. The dual treatment will allow us to model an inductor L. Any numerical scheme based on a finite time discretization Δt is associated with some error. In schemes that are based on representing lumped components by TL segments the transit time τ is associated with the sampling interval Δt and the associated errors may be interpreted in terms of parasitic inductance (when modeling C) and parasitic capacitance (when modeling L).

It is easy to see how a short segment of a TL can be used to model a capacitance. Consider the TL segment shown in Fig. 3.1, assume that the end BB is an open circuit and calculate the input impedance Z_{in} at the port AA. From (3.12) setting $Z_1 \rightarrow \infty$ we obtain $Z_{in} = Z/j \tan(\beta \ell)$ and if the length of the segment is much shorter than the wavelength then $\tan(\beta \ell) \approx \beta \ell$. It follows therefore that at low frequencies ($\ell \ll \lambda$)

$$Z_{in} \approx \sqrt{\frac{L_d \ell}{C_d \ell}}/(j\omega\sqrt{L_d C_d}\ell) = 1/j\omega(C_d \ell) \qquad (3.13)$$

We see therefore that provided the TL segment is electrically short it can be viewed as a small capacitance. Similarly, a segment short circuited at BB looks like an inductance. We now look in more detail at these models.

It turns out that there are two possible models of a capacitor: a two-port model (a "link" line) and a one-port model (a "stub" line). We look first at the "stub" model of a capacitor. The basic model is a TL segment open circuit at one end with a round-trip time equal to Δt (hence in this case the transit time τ is equal to $\Delta t/2$) and length ℓ as shown in Fig. 3.2(a). What should be the characteristic impedance Z_c of this stub so that it models the desired capacitance C? The answer is obtained directly from Eqs. (3.10) by setting $\tau = \Delta t/2$,

$$Z_c = \frac{\Delta t/2}{C} = \frac{\Delta t}{2C} \qquad (3.14)$$

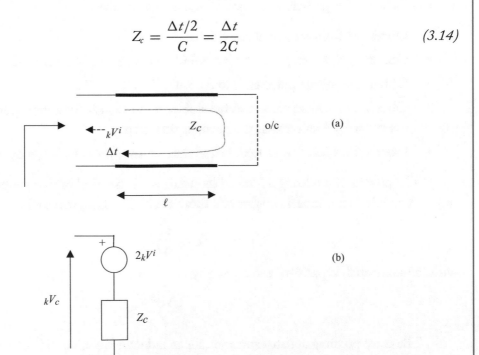

FIGURE 3.2: An open-circuit stub representing a capacitor C (a) and its Thevenin equivalent circuit (b)

The associated modeling error, in this case a "parasitic" inductance, is obtained in a similar way.

$$L_{\text{err}} = \frac{\Delta t}{2} Z_c = \frac{(\Delta t)^2}{4C} \qquad (3.15)$$

We observe that the error in (3.15) can be made smaller by reducing Δt. This is reasonable as the finer the time sampling the better the model should describe C. The signal at the port of the stub is updated at time intervals Δt (at times $k\Delta t$ where k is an integer)—in effect Δt is the time discretization length. A voltage pulse V^r reflected from the port of the stub at time $k\Delta t$ reaches the open-circuit termination after time $\Delta t/2$, it is reflected with the same sign V^r and becomes incident at the port at the next time step $(k+1)\Delta t$. The capacitive nature of the stub is embodied in this procedure, i.e., the voltage pulse incident at the port at time $(k+1)\Delta t$ is equal to the pulse reflected from the same port at the previous time step $k\Delta t$,

$$_{k+1}V^i = {}_kV^r \qquad (3.16)$$

Looking into the port of the stub from the rest of the network at time $k\Delta t$ one can derive a Thevenin equivalent circuit valid for a sampling interval as shown in Fig. 3.2(b). The discrete model of a lumped capacitance component operates as follows:

- Obtain V^i from the initial conditions.
- Obtain V_c by solving the circuit in which the capacitor is connected.
- Obtain the voltage reflected into the stub $V^r = V_c - V^i$.
- Obtain the incident voltage at the stub port at the next time step which in this case is the voltage V^r reflected at the previous time step.
- Proceed to repeat this procedure and thus advance the calculation by one time step.

The process of modeling a lumped inductance L is the dual of that adopted for a capacitance. A stub inductor model consists of a short-circuited TL segment where

$$Z_L = \frac{2L}{\Delta t} \qquad (3.17)$$

with the associated "capacitive" error given by

$$C_e = \frac{(\Delta t)^2}{4L} \qquad (3.18)$$

The corresponding update expression for an inductor is similar to (3.16) but with a minus sign to recognize that the far end of an inductive stub is a short circuit, i.e.,

$$_{k+1}V^i = -{}_kV^r \qquad (3.19)$$

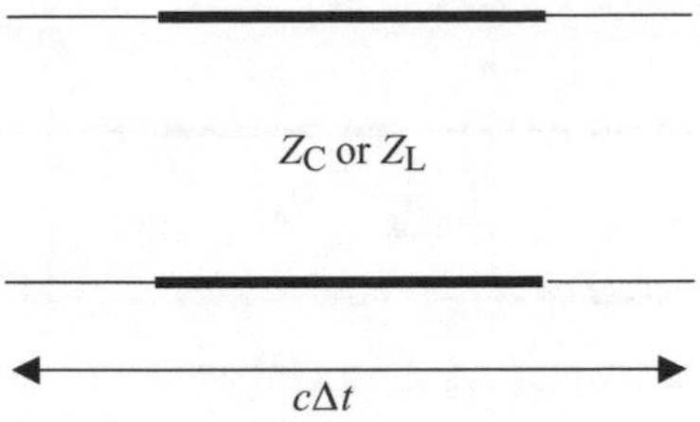

Z_C or Z_L

$c\Delta t$

$$Z_C = \frac{\Delta t}{C}, \; L_e = \frac{(\Delta t)^2}{C}$$

$$Z_L = \frac{L}{\Delta t}, \; C_e = \frac{(\Delta t)^2}{L}$$

FIGURE 3.3: A link-line model of a capacitor (Z_C) and an inductor (Z_L)

Link models of capacitors and inductors are line segments open at both ends with single transit time Δt and characteristic impedance given Z_C, Z_L for capacitors and inductors, respectively as shown in Fig. 3.3 where also the associated errors L_e, C_e are given.

The modeler is free to choose stub or link models depending on the nature of the problem and convenience. Errors however combine differently and some choices may be slightly better than others. Provided the modeling error is kept low accuracy will be acceptable. I also stress here that since we have a physical interpretation of errors ("stray" component) it is easier for the modeler to assess their impact and therefore control them. As an illustration, if we are modeling a capacitor we know that a certain stray inductance L_e is associated with the model. Provided L_e is small it may be viewed as the inevitable inductance associated with the leads of an actual capacitor and therefore perfectly legitimate as a representation of a real capacitor. In some cases where we have inductors and capacitors in the same circuit the modeling error C_e associated with L may be subtracted from C and thus minimize errors further. A careful consideration of modeling errors, reductions in Δt, and adjustments along the lines mentioned above can maintain a high accuracy in calculations. A fuller description of errors may be found in [4].

One can regard the TLM models of lumped components L and C as a more general class of Discrete TLM Transforms which may be applied like Laplace Transforms (LT) to solve circuits or general integro-differential equations. Whilst the LT transforms to the s-domain, the Discrete TLM Transform transforms directly to the discrete time-domain thus offering a powerful and elegant solution [4, 9].

3.3   ONE-DIMENSIONAL TLM MODELS

A good physical insight may be gained by the study of one-dimensional (1D) problems where variation in only one coordinate is allowed. Thus the mathematical difficulties are minimized

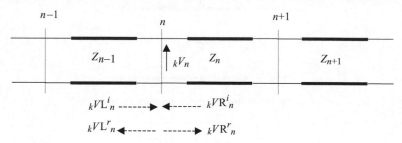

FIGURE 3.4: A cascade of segments, part of a transmission line, showing notation for incident and reflected voltage pulses

and a clearer physical picture may thus be gained. Such problems are also useful in illustrating the TLM algorithm in a clear way so I devote this section to show how such problems are tackled with TLM.

A 1D problem is modeled by a cascade of TL segments as shown in Fig. 3.4. The parameters of each segment are chosen to represent the physical system being modeled, e.g., if a capacitor is modeled then the segment has a characteristic impedance given by $Z = \Delta t / C$ and Δt is chosen small enough to make the modeling error (stray inductance) negligible. The novice modeler will be concerned about the choice of Δt and rightly so. Various related factors impinge on its final choice. It must be small enough to minimize modeling errors as indicated above, it must be much smaller than the period of the highest frequency of interest and in the case of transients it must be much smaller than the shortest transition time. The latter two are necessary in order to allow a proper study of the phenomenon. In the case of spatially extensive circuits the choice of Δt implies a choice of the spatial discretization length Δl through the relationship $\Delta l = u \Delta t$ where u is the velocity of propagation of electrical disturbances along the circuit. In such cases the spatial discretization length must be much smaller than the wavelength of the highest frequency of interest. A legitimate question is "How much smaller?" Well, it depends on the desired accuracy!

Considering that a smaller time step means a much larger computation, there is strong incentive to keep Δt as high as possible consistent with an acceptable accuracy. Most people are happy with a discretization length that is smaller than a tenth of the shortest wavelength (accuracy of a few percent). However, a finer resolution may be necessary in some problems. The same time step must be employed throughout the model to maintain "synchronism", i.e., the exchange of pulses at the boundary between adjacent segments must take place at the same moment. In cases where the study above has revealed a number of possible time steps meeting the necessary conditions at different parts of the model, then the shortest time step is chosen and imposed throughout the model. It is evident therefore that the way that a problem is discretized in time and in space requires an understanding of the features of the TLM model and also of the inherent physics. There is no "single" correct answer—many valid alternatives are possible. As your confidence in modeling increases you will be able to make inspired choices in the way you construct the model (mix of stubs and links) and therefore set a time step that is a fair compromise between accuracy and computational efficiency.

Let us now return to Fig. 3.4 and examine how to operate the TLM algorithm. Three segments are shown for simplicity. The junction between adjacent segments I describe as a node and I have labeled nodes $n - 1$, n, and $n + 1$. The transit time across each segment is the same throughout the model (synchronism), however I have allowed for the characteristic impedance to be different in each segment (Z_n, etc.). Standing as an observer at any particular node I experience pulses coming towards me from the left and right (incident) and also pulses moving away from me to the left and to the right (reflected). I need an efficient labeling scheme for these pulses because throughout the computation I will need to keep a close account of pulses at every node. The labeling scheme is explained for pulse $_kVL_n^i$ shown in Fig. 3.4. This pulse is incident (i) on node n from the left (L) at time $k\Delta t$ (k). All the other pulses are labeled following the same principles. At time step k an observer at node n will be able to look left and replace what he or she sees by a Thevenin equivalent circuit and also look right and do exactly the same. The conditions at node n at time step k are therefore as depicted in Fig. 3.5. The voltage at the node is therefore given by (Millman's Theorem).

$$_kV_n = \frac{\dfrac{2_kVL_n^i}{Z_{n-1}} + \dfrac{2_kVR_n^i}{Z_n}}{\dfrac{1}{Z_{n-1}} + \dfrac{1}{Z_n}} \tag{3.20}$$

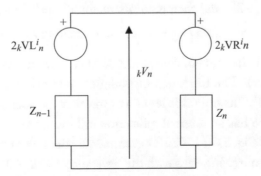

FIGURE 3.5: Thevenin equivalent circuit at node n and at time k

The voltage pulses reflected to the right and to the left can now be directly calculated since the total voltage anywhere on a TL is the sum of incident and reflected voltages, hence,

$$_kVL_n^r = {_k}V_n - {_k}VL_n^i \qquad _kVR_n^r = {_k}V_n - {_k}VR_n^i \qquad (3.21)$$

Therefore, given the incident voltages at a particular instance we can calculate the reflected voltages from (3.20) and (3.21) in a process which in TLM is described as *scattering*. The question now is how do we get the incident voltages, how do we start? Naturally, the first time we do this calculation ($k = 1$) we start with the *initial conditions* that provide the value of all incident voltage pulses. But what happens after at the next time step ($k = 2$)? We have already used the initial conditions and need to generate the incident voltage at the next time step from within our solution procedure. This is particularly simple and it is described as the *connection* process in TLM. Examining the topology shown in Fig. 3.4 we see that the pulse reflected at time k from node n and travelling to the left becomes incident on node $n - 1$ from the right at time $k + 1$, i.e.,

$$_{k+1}VR_{n-1}^i = {_k}VL_n^r \qquad (3.22)$$

By the same logic the incident pulse from the left at node n and at time $k + 1$ is

$$_{k+1}VL_n^i = {_k}VR_{n-1}^r \qquad (3.23)$$

Similar expressions apply for all the new incident voltages to all nodes. In summary, the solution proceeds as follows:

- Using the *initial conditions* we obtain the incident voltages on all the nodes at the start of the calculation.

- We perform *scattering* on all the nodes to calculate the reflected voltages [Eq. (3.21)].

- We perform the *connection* to calculate the incident voltages at the next time step using Eqs. (3.22) and (3.23) and their equivalents for all nodes and directions of incidence.

- We repeat the process *scattering–connection* for as long as required.

I have glossed over the issue of source and load conditions (what is known as *boundary conditions* in mathematics). The treatment of boundary conditions is very similar to what we have already discussed. To illustrate this let us assume that the source conditions are as shown in Fig. 3.6(a)—the source has an internal resistance and inductance. The conditions at node 1 are therefore as shown in Fig. 3.6(b). The Thevenin equivalent circuit looking right is as before. Looking left we see the source in series with the resistance and a stub model of the inductance. I have used a stub in this case as it is more convenient! I label pulses coming from the stub as $_kV_L^i$. Looking into the stub I can replace it by its Thevenin equivalent as shown in Fig. 3.6(c).

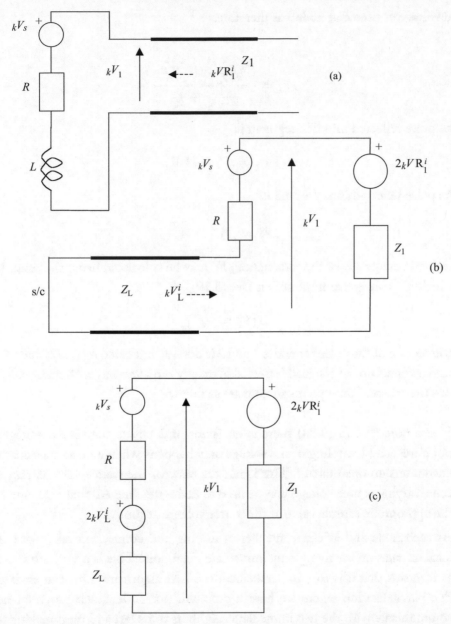

FIGURE 3.6: Connection at the source end of a transmission line (a), TLM equivalent (b), and conversion to the Thevenin equivalent circuit (c)

The total voltage at time k at node 1 is therefore,

$$_kV_1 = \frac{\frac{_kV_s + 2_kV_L^i}{R+Z_L} + \frac{2_kVR_1^i}{Z_1}}{\frac{1}{R+Z_L} + \frac{1}{Z_1}} \quad (3.24)$$

The pulse reflected into line segment is

$$_kVR_1^r = {_kV_1} - {_kVR_1^i} \quad (3.25)$$

The pulse reflected into the stub is

$$_kV_L^r = {_kV_L} - {_kV_L^i} \quad (3.26)$$

where the total voltage across the inductance $_kV_L$ may be calculated using the result in (3.24). The new incident voltage from the stub is [see (3.19)],

$$_{k+1}V_L^i = -_kV_L^r \quad (3.27)$$

and the connection of the pulses at nodes 1 and 2 is done as indicated by (3.22) and (3.23). The other boundary condition at the load is treated in exactly the same way with minor adjustments, i.e., remove the source. Some points to note are as follows:

- Please note that in (3.24) there is no factor of 2 before the source voltage. This is not a mistake! Doubling of the voltage only happens when a pulse travelling on a TL encounters an open circuit. This is not the case for the source voltage. Any specified time-varying source voltage may be used to excite the line. All that is required is to use the appropriate sample value of the source voltage at time k.

- At each node and at each time step scattering and connection take place. This is a local calculation involving only immediate neighbors. This is a very attractive feature as it means that it is easy to parallelize the TLM algorithm. One can envisage a very fine parallelization where we have a processor per node. Each processor need only communicate with the two immediate neighbors to its left and right making for a very efficient operation. The physical justification for the local nature of the calculation at each node comes from the fact that in the time duration of a time step Δt we can only "see" a distance $u\Delta t$ around us (u is the speed of signal propagation).

You will be surprised how many useful problems you can solve with the techniques described so far. A particular example is interconnect problems to assess the impact of discontinuities. These problems can be easily and quickly solved. In addition, problems other than electrical problems can be solved, e.g., thermal problems, using similar procedures [4].

However, most practical problems require a formulation at least in two dimensions and there are many which can be dealt with adequately only in three dimensions. We therefore need to establish such models in the next few chapters. But the philosophy of modeling and the basic techniques remain the same—only complexity is added.

CHAPTER 4

Two-Dimensional TLM Models

4.1 BASIC CONCEPTS

I have introduced 1D TLM based on the isomorphism between transmission line and field equations. It is possible to do the same for two-dimensional (2D) TLM but I prefer to show you another way that connects with a powerful concept in EM theory. Huygens in 1690 [10] stated that a wavefront propagates by a mechanism whereby each point on the wavefront acts as an isotropic spherical radiator and that the superposition of all these elementary point radiators forms a new wavefront and so on. We can introduce 2D TLM by analogy to Huygens principle.

Consider two intersecting TLs as shown in Fig. 4.1(a) each of the same length and of characteristic impedance Z. We launch a pulse equal to 1 V on port 1 and proceed to calculate how this pulse will scatter when it reaches the junction between the lines. At the junction, the pulse sees three identical lines in parallel and therefore encounters an impedance R equal to $Z/3$. The reflection coefficient is therefore equal to $(R - Z)/(R + Z) = -0.5$ and the transmission coefficient $2R/(R + Z) = 0.5$. We show the two lines in a simpler form as a one-line diagram in Fig. 4.1(b). The original pulse of 1 V has now generated four new pulses, three transmitted and one reflected as shown in Fig. 4.1(b), i.e., we now have a spherical wave of amplitude 0.5 V in each direction. We see here a way to represent Huygens principle in a discrete way. If we imagine the entire problem space to be populated by a grid of TLs [replicate Fig. 4.1(b) in each direction] then we have a view of propagation exactly as Huygens envisaged, i.e., each pulse reflected from the node impinges on the adjacent node and sets up a spherical wave. The pulses associated with this wave become incident on adjacent nodes to set up more spherical waves, the entire grid of lines being the modeling medium on which the pulses propagate and scatter. The evolution of this phenomenon is by analogy an image of the EM resulting from the original disturbance. The first few time steps following the excitation of the mesh by four equal pulses are shown in Fig. 4.2. If we can for a moment use poetic licence, watching propagation of pulses in this figure is like watching the propagation of ripples on the surface of calm water following the dropping of a pebble—electromagnetics can be that simple!

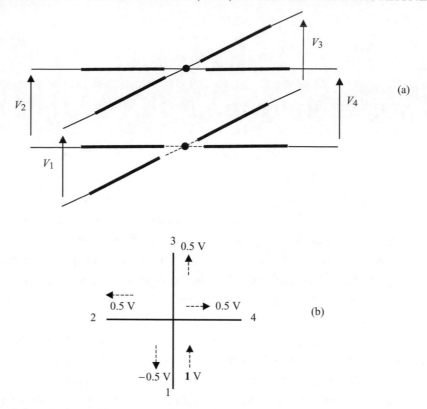

FIGURE 4.1: A node formed by the intersection of two transmission lines (a) and schematic representation showing scattering of incident pulse of 1 V on port 1 (b)

Naturally, we need to work out details of the exact topology of the TL network that makes up the TLM model, the TL parameters corresponding to different media and how electric and magnetic fields map onto the TLM model. We will do this in the next sections.

To set the scene, let us examine the options available to us when we wish to model EM propagation in a block of space of dimensions $\Delta x \Delta y \Delta z$ as shown in Fig. 4.3(a). We can envisage two 2D TL configurations that fit into the block (cell) of Fig. 4.3(a). First, we have the "series" arrangement of TLs shown in Fig. 4.3(b) where it is expected that port voltages V_1 and V_3 will be associated with the electric field component E_x (since these voltages are polarized in the x-direction) and that V_2, V_4 will be associated with E_y. Similarly, the circulating current I_z will be associated with the magnetic field component H_z. Thus this so-called *series node* models field components E_x, E_y, H_z, i.e., TE modes. The second option is the structure shown in Fig. 4.3(c) known as the *shunt node*. Here it is expected is that voltages V_1–V_4 will be associated with the electric field component E_z and that V_1 and V_3, since they generate a current on the y–z plane, will be associated with the magnetic field component H_x. Similarly, V_2 and V_4 will be

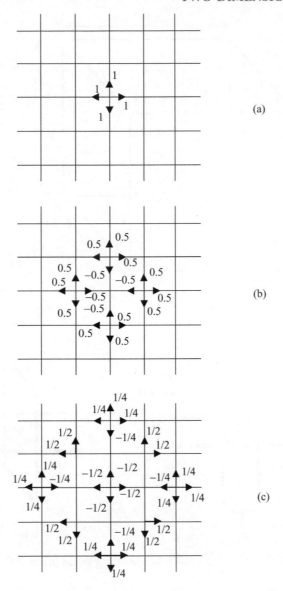

FIGURE 4.2: Initial symmetric excitation in a 2D mesh (a), and situation after one (b) and two (c) time steps

associated with H_y. The shunt node therefore models E_z, H_x, H_y, i.e., TM modes. Using these two structures therefore allows modeling of all field components. The modeling is physically consistent in that electric fields are associated with voltages and magnetic fields with currents. If one is prepared to dispense with this more natural choice then it is possible to use one of the two nodes to model both TE and TM modes. It means, however, that voltages will be used as

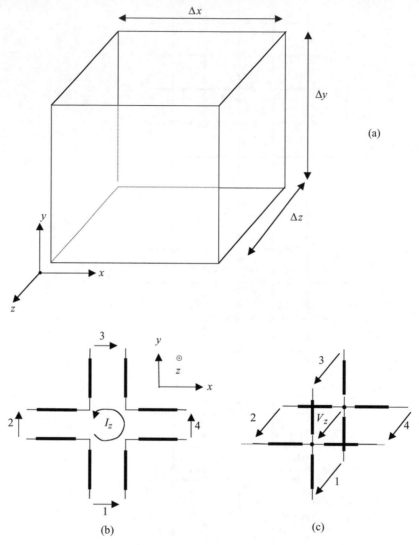

FIGURE 4.3: A cuboid-shaped cell (a), and corresponding series (b), and shunt (c) TLM nodes

equivalents for magnetic fields, etc. (duality in electromagnetics). The advantage of this fact is that only one structure need be implemented in software and deals with both modes [11–13].

I will introduce the basics of each type of node in the sections that follow.

4.2   MODEL BUILDING WITH THE SERIES NODE

The basic element of a mesh made out of series nodes is shown in Fig. 4.3(b). We assume for simplicity that $\Delta x = \Delta y = \Delta \ell$ and that all TLs have the same characteristic impedance Z_{TL}.

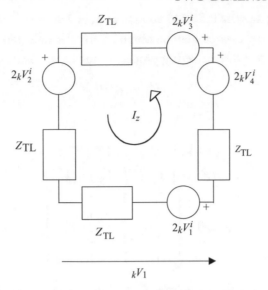

FIGURE 4.4: Thevenin equivalent circuit of a series node

The choice of the value of $\Delta\ell$ is dependent on the spatial resolution desired and the shortest wavelength of interest ($\Delta\ell < \lambda/10$ for accuracy). Z_{TL} depends on the parameters of the medium in which propagation takes place. We will discuss these matters in more detail but let us first sketch out the basic operation of the TLM model.

At each time step k and each node there will be four incident pulses which after scattering at the node generate four scattered (or reflected) pulses. These propagate out of each node to become incident on adjacent nodes at the next time step $k + 1$ and the process repeats. That is all! We now need to add the maths to the *scattering* and *connection* processes I have outlined.

An observer at node (x, y, z) can replace what he or she sees by the Thevenin equivalent to obtain the circuit shown in Fig. 4.4. The loop current at time step k is then

$$_kI_z = \frac{2_kV_1^i + 2_kV_4^i - 2_kV_3^i - 2_kV_2^i}{4Z_{TL}} \qquad (4.1)$$

where all the quantities are evaluated at node (x, y, z).

The total voltage across port 1 is then

$$_kV_1 = 2_kV_1^i - {_kI_z}Z_{TL}$$

and the reflected voltage at port 1 is

$$\begin{aligned}_kV_1^r &= {_kV_1} - {_kV_1^i} = {_kV_1^i} - {_kI}Z_{TL} \\ &= 0.5({_kV_1^i} + {_kV_2^i} + {_kV_3^i} - {_kV_4^i})\end{aligned} \qquad (4.2)$$

This expression gives the reflected voltage at port 1 in terms of the incident voltages at the four ports. Similar expressions may be obtained for the reflected voltage at the remaining three ports. We can express the scattering process in terms of a scattering matrix S,

$$_kV^r = S_k V^i \tag{4.3}$$

where

$$_kV^r = [_kV_1^r \,\, _kV_2^r \,\, _kV_3^r \,\, _kV_4^r]^T$$
$$_kV^i = [_kV_1^i \,\, _kV_2^i \,\, _kV_3^i \,\, _kV_4^i]^T$$
$$S = 0.5 \begin{bmatrix} 1 & 1 & 1 & -1 \\ 1 & 1 & -1 & 1 \\ 1 & -1 & 1 & 1 \\ -1 & 1 & 1 & 1 \end{bmatrix} \tag{4.4}$$

where superscript T stands for transpose. Equations (4.3) and (4.4) embody the *scattering* process in each node. Scattering is particularly simple as it involves a simple arithmetic averaging.

We move now to consider the connection process whereby new incident pulses are obtained to allow the calculation to proceed to the next time step $k + 1$.

We show in Fig. 4.5 one node and its immediate neighbors. As I have pointed out before, the connection is nothing more than a recognition of the fact that a pulse reflected from the

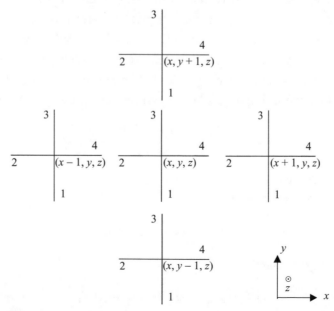

FIGURE 4.5: A cluster of nodes to illustrate the connection process

center of the node (x, y, z) at time step k and travelling say towards port 1 will become incident at time step $k + 1$ at node $(x, y - 1, z)$ at its port 3 (see Fig. 4.5). Similar statements may be made for all other new incident voltages. In mathematical form we can state in a formal way that

$$_{k+1}V^i = C_k V^r \tag{4.5}$$

where C is a *connection* matrix. It is not necessary to express C in an explicit form—connection is simply an exchange of pulses with immediate neighbors as shown below.

$$\begin{aligned}
_{k+1}V_1^i(x, y, z) &= {}_k V_3^r(x, y - 1, z) \\
_{k+1}V_2^i(x, y, z) &= {}_k V_4^r(x - 1, y, z) \\
_{k+1}V_3^i(x, y, z) &= {}_k V_1^r(x, y + 1, z) \\
_{k+1}V_4^i(x, y, z) &= {}_k V_2^r(x + 1, y, z)
\end{aligned} \tag{4.6}$$

To summarize, computation involves the following steps:

- Using the initial conditions determine all incident voltages on all nodes at $k = 1$.
- Scatter at all nodes [Eqs. (4.3) and (4.4)].
- Obtain incident voltages at $k + 1$ by implementing the connection process at all nodes [Eq. (4.6)].
- Scatter again at $k + 1$ and continue for as long as desired.

The only other complication is to deal with boundary conditions something which we will discuss after we have dealt with both the series and shunt nodes. But just in case you are concerned that this may be difficult I show here how to deal with a simple boundary condition—a mesh terminated by a perfect electric conductor (PEC). This could be the surface of a conducting box, for example. The situation is depicted in Fig. 4.6 where a node is shown with PEC boundary on port 4. Connection for ports 1–3 is done exactly as implied by Eqs. (4.6). However, port 4 has no immediate neighbors other than the PEC boundary so the connection process here must recognize this. A pulse reflected from this node (x, y, z) at time step k travelling out of port 4 will encounter a short circuit (PEC boundary) and will be reflected with an opposite sign to become incident on the same node and port at time step $k + 1$, i.e.,

$$_{k+1}V_4^i(x, y, z) = -_k V_4^r(x, y, z) \tag{4.7}$$

Thus, the presence of the conducting boundary is accounted for very simply at the connection phase of the algorithm.

We must now turn our attention to how we excite the mesh (connect sources) and obtain the output (electric and magnetic fields).

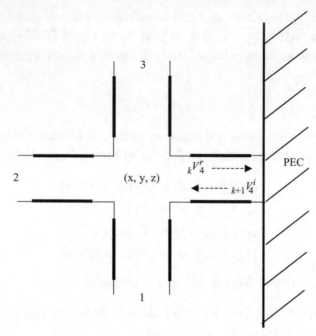

FIGURE 4.6: A series node terminated by a perfect electric conductor (PEC)

H_z is related to the current in Eq. (4.1) through Ampere's Law,

$$_kH_z = \frac{_kI_z}{\Delta\ell} = \frac{_kV_1^i - {_kV_2^i} - {_kV_3^i} + {_kV_4^i}}{2\Delta\ell\, Z_{\text{TL}}} \qquad (4.8)$$

where all the quantities are evaluated at the node in question.

The two electric field components are similarly given by

$$_kE_x = -\frac{_kV_1^i + {_kV_3^i}}{\Delta\ell} \qquad (4.9)$$

$$_kE_y = -\frac{_kV_2^i + {_kV_4^i}}{\Delta\ell} \qquad (4.10)$$

From these expressions it is clear that if for example we wish to impose an electric field of magnitude E_0 in the x-direction, then we need to apply to this node the following pulses, $_kV_1^i = {_kV_3^i} = -E_0\Delta\ell/2$. This arrangement excites no other field component. Similarly, to excite $H_z = H_0$ we must apply the following pulses

$$\begin{aligned} _kV_1^i &= {_kV_4^i} = H_0\Delta\ell\, Z_{\text{TL}}/2 \\ _kV_3^i &= {_kV_2^i} = -H_0\Delta\ell\, Z_{\text{TL}}/2 \end{aligned} \qquad (4.11)$$

Excitation can be a continuous time-dependent function, a Gaussian pulse, or more often in TLM an impulse of duration Δt. In the latter case the impulse response of the system is obtained. The frequency response may in turn be derived by a Fourier Transform of the impulse response. However, you should be aware that the broadband response is of acceptable accuracy for frequencies $f \leq 1/(10\Delta t)$ [see discussion leading to (2.24)].

So far, I have not said much on the medium in which propagation takes place. This could be free space (air) or another dielectric or magnetic medium, or a mixture (inhomogeneous medium). The parameters of the TLM model (Z_{TL}, Δt) must be chosen subject to two constraints:

First, the total capacitance and inductance represented by the model must accord with the electric and magnetic properties of the block of space (the cell) represented by the node, i.e., $\Delta x \Delta y \Delta z = (\Delta \ell)^3$. From Fig. 4.3(a) we see that the x-directed capacitance is

$$C_x = \varepsilon \frac{\Delta y \Delta z}{\Delta x} \tag{4.12}$$

where ε is the dielectric permittivity of the medium. This capacitance must be represented in the model by the capacitance of the line joining ports 1 and 3 of the series node. Similarly,

$$C_y = \varepsilon \frac{\Delta x \Delta z}{\Delta y} \tag{4.13}$$

must be the total capacitance of the TL joining ports 2 and 4 in the model. To find the inductance L_x associated with x-directed line currents we employ Ampere's law.

$$L_x \equiv \frac{\Phi_z}{I_x} = \frac{\mu H_z \Delta x \Delta y}{I_x} = \mu \frac{\Delta x \Delta y}{\Delta z} \tag{4.14}$$

where the first part of (4.8) is used to substitute for H_z/I_x. Similarly, we can calculate L_y,

$$L_y = \mu \frac{\Delta x \Delta y}{\Delta z} \tag{4.15}$$

Returning to our choice of $\Delta x = \Delta y = \Delta z = \Delta \ell$ Eqs. (4.12)–(4.15) simplify to,

$$L = \mu \Delta \ell \qquad C = \varepsilon \Delta \ell \tag{4.16}$$

In a correctly constituted model voltage pulses (electric field) must experience C and the current pulses (magnetic field) must experience L in (4.16).

Second, as we have already pointed out in previous chapters, it is important that synchronism is maintained throughout the TLM mesh (same time step). This becomes an issue when modeling inhomogeneous materials.

I now illustrate how mesh parameters are chosen for two cases: *propagation in free space* (ε_0, μ_0) *and in a magnetic medium* (ε_0, μ).

When propagation takes place in free space and provided a regular mesh is employed (same $\Delta\ell$ throughout the mesh) the issue of synchronism does not arise—$\Delta\ell$ is chosen to be smaller than a tenth of the wavelength and also to accommodate geometrical features of importance. We make the following choice

$$u_{\mathrm{TL}} = \frac{\Delta\ell}{\Delta t} = \sqrt{2}c$$

$$Z_{\mathrm{TL}} = \frac{1}{\sqrt{2}}\eta$$

(4.17)

where c is the speed of light $(= 1/\sqrt{\mu_0\varepsilon_0})$ and η is the intrinsic impedance of free space $(= \sqrt{\mu_0/\varepsilon_0})$. We will now show that this choice correctly models free space. We simply need to demonstrate that we model the correct amounts of L and C as calculated in (4.16).

From Eq. (3.10) we know that the inductance modeled TL is

$$L = Z_{\mathrm{TL}}\Delta t = \frac{\eta}{\sqrt{2}}\frac{\Delta\ell}{\sqrt{2}c} = \mu_0\Delta\ell/2$$

However, the current loop current encounters the inductance L of the vertically directed line and also another L due to the horizontally directed line. In total, therefore, the current correctly encounters an inductance $2L = \mu_0\Delta\ell$. Similarly, each electric field component sees a capacitance

$$C = \frac{\Delta t}{Z_{\mathrm{TL}}} = \frac{\Delta\ell}{\sqrt{2}c}\frac{\sqrt{2}}{\eta} = \eta_0\Delta\ell$$

as required. These expressions are summarized in Fig. 4.7.

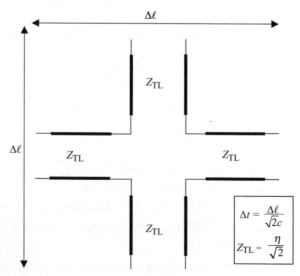

FIGURE 4.7: Parameters of a series node representing free space

We now turn our attention to the modeling of a medium (ε_0, μ). If this was the only medium present in the problem space then a minor adjustment to the conditions in Fig. 4.7 would suffice. Simply, we would need to replace c by $c/\sqrt{\mu_r}$ and η by $\eta\sqrt{\mu_r}$. However, most engineering problems are inhomogeneous—there is free space and also some other material (magnetic in the present case). If we do what I have suggested above, then we will end up with parts of the mesh representing free space having a different time step to those representing the magnetic material, i.e., we will lose synchronism. It may be argued that in spite of the fact that the velocity of propagation in the two media is different we need not have a different time step provided we have a different $\Delta\ell$ in the two media. This simply removes the loss of synchronism but it introduces another problem in its place. We have lost one-to-one correspondence in space—each port has no clear unique neighbor. We may end up with a port at the interface between the two media having say 1.35 neighbors! Clearly this will create all kinds of problems. We need to retain the same time step throughout the problem and also the same space discretization. I propose that we pretend that the entire space is free space and set Δt, $\Delta\ell$ accordingly. Then, in places where $\mu > \mu_0$ we will calculate how much extra inductance we need to add and we will add it in some other way.

Therefore, as in the case of free space the choice of parameters is as shown in Eqs. (4.17). This then implies that the link lines model a total capacitance and inductance

$$
\begin{aligned}
C_{\text{link}} &= \varepsilon_0 \Delta\ell \\
L_{\text{link}} &= \mu_0 \Delta\ell
\end{aligned}
\tag{4.18}
$$

I have added the subscript "link" to emphasize that these are the values modeled by the link lines in the series node model. The amount of capacitance modeled is as required, but we have a deficit in the amount of inductance introduced by the link lines. This deficit is

$$
L_s = \mu\Delta\ell - L_{\text{link}} = (\mu_r - 1)\mu_0\Delta\ell
\tag{4.19}
$$

How can we introduce this missing inductance to the model without affecting synchronism (same Δt) and connectivity (same $\Delta\ell$)? The answer is that we will break the loop and load the current with an additional inductance given by (4.19). This inductance L_s may then be modeled by a short-circuit stub as explained in Section 3.2. All the details are shown in Fig. 4.8. Note that provided free space is chosen as the background medium, which sets the time step, then L_s is always positive (we need to *add* inductance to the model). It would be very inconvenient if we had to remove inductance (negative L_s) as this may lead to instabilities. So make sure that you construct your models in such a way that you always have stubs representing positive parameters!

You may have noticed that Figs. 4.7 and 4.8 look different. The scattering matrix in (4.4) cannot be used for the node in Fig. 4.8. We need to derive a new S matrix but this is not difficult.

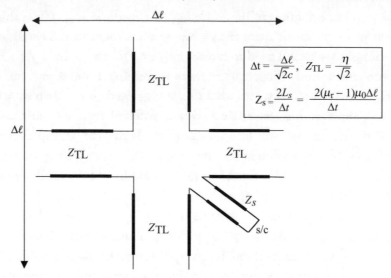

$$\Delta t = \frac{\Delta \ell}{\sqrt{2}c}, \quad Z_{TL} = \frac{\eta}{\sqrt{2}}$$

$$Z_s = \frac{2L_s}{\Delta t} = \frac{2(\mu_r - 1)\mu_0 \Delta \ell}{\Delta t}$$

FIGURE 4.8: Parameters of a series node representing medium (ε_0, μ)

The Thevenin equivalent circuit of the circuit in Fig. 4.8 is as shown in Fig. 4.9 where $_kV_5^i$ is the pulse incident at the node from the stub. The loop current is

$$_kI_z = \frac{2_kV_1^i - 2_kV_2^i - 2_kV_3^i + 2_kV_4^i + 2_kV_5^i}{4Z_{TL} + Z_s} \tag{4.20}$$

This equation is the equivalent of (4.1) for the case without the stub. The calculation of the total voltage across each port and of the scattered voltages proceeds along the same lines

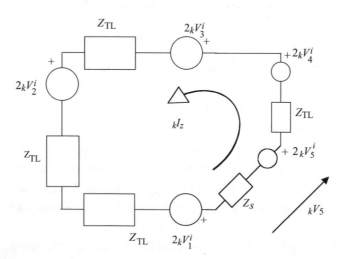

FIGURE 4.9: Thevenin equivalent circuit for the node in Fig. 4.8

but with the extra terms representing the stub. The scattering matrix is now

$$S = \frac{1}{Z} \begin{bmatrix} Z-2 & 2 & 2 & -2 & -2 \\ 2 & Z-2 & -2 & 2 & 2 \\ 2 & -2 & Z-2 & 2 & 2 \\ -2 & 2 & 2 & Z-2 & -2 \\ -2Z_s/Z_{TL} & 2Z_s/Z_{TL} & 2Z_s/Z_{TL} & -2Z_s/Z_{TL} & 4-Z_S/Z_{TL} \end{bmatrix} \quad (4.21)$$

where $Z = 4 + Z_s/Z_{TL}$. The approach just described, to add stubs in order to account for nonuniformities, is used in many situations to facilitate computation.

A more extensive treatment of the series node with further fundamental considerations, inclusion of losses, etc. may be found in [4].

4.3   MODEL BUILDING WITH THE SHUNT NODE

A treatment analogous to that for the series node is also necessary to implement scattering and connection at the shunt node shown in Fig. 4.3(c). We assume again voltage pulses incident at a node and replace what we see coming from each TL by its Thevenin equivalent. Conditions at the center of the node are thus as described in Fig. 4.10 and it is easy to show that the total voltage V_z is given by the formula,

$$_kV_Z = \frac{1}{2}\left[_kV_1^i + _kV_2^i + _kV_3^i + _kV_4^i \right] \quad (4.22)$$

Following the same procedure as for Eqs. (4.1) and (4.2) we can obtain the reflected voltage from port 1

$$_kV_1^r = _kV_Z - _kV_1^i = \frac{1}{2}\left[-_kV_1^i + _kV_2^i + _kV_3^i + _kV_4^i \right] \quad (4.23)$$

This equation contains the elements of the first row of the scattering matrix for this node. The other rows are similarly obtained by calculating the scattering voltages from the remaining

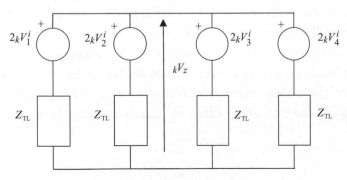

FIGURE 4.10: Thevenin equivalent circuit for the shunt node

three ports to give

$$S = \frac{1}{2} \begin{bmatrix} -1 & 1 & 1 & 1 \\ 1 & -1 & 1 & 1 \\ 1 & 1 & -1 & 1 \\ 1 & 1 & 1 & -1 \end{bmatrix} \tag{4.24}$$

The connection process is identical to that for the series node. As already pointed out this node represents TM modes and the corresponding field components are obtained as follows,

$$E_z = -\frac{V_z}{\Delta z} = -\frac{1}{2\Delta z}\left(V_1^i + V_2^i + V_3^i + V_4^i\right) \tag{4.25}$$

$$H_y = \frac{I_x}{\Delta y} = \frac{V_2^i - V_4^i}{Z_{TL}\Delta y} \qquad H_x = -\frac{I_y}{\Delta x} = \frac{V_3^i - V_1^i}{Z_{TL}\Delta x} \tag{4.26}$$

where in Eq. (4.26) we have associated the y-component of the magnetic field with x-directed current, and the x-component with y-directed current. Excitation proceeds along similar lines as for the series node.

The choice of parameters to model a particular medium follows similar principles as for the series node. I illustrate here by an example the modeling of a medium with parameters ε, μ_0. Since, inevitably, in most problems free space is also present, I choose the background medium to be free space and in the areas with a higher dielectric permittivity I insert stubs to represent an $\varepsilon > \varepsilon_0$. I choose

$$u_{TL} = \sqrt{2}\,\frac{1}{\sqrt{\varepsilon_0 \mu_0}} \tag{4.27}$$

$$Z_{TL} = \sqrt{2}\sqrt{\frac{\mu_0}{\varepsilon_0}} \tag{4.28}$$

Thus, the modeled inductance is as required $L = Z_{TL}\Delta t = \mu_0\Delta\ell$ where I have assumed for simplicity that all dimensions are $\Delta\ell$. The modeled capacitance is $C = \Delta t / Z_{TL} = 0.5\varepsilon_0\Delta\ell$ and since the electric field component experiences this capacitance twice (once for each of the TLs) the total modeled capacitance is $\varepsilon_0\Delta\ell$. This would be fine in the free-space regions. In our case we have a deficit in modeled capacitance that is $C_s = (\varepsilon_r - 1)\,\varepsilon_0\Delta\ell$. This we can introduce into the model in the form of a stub where the round-trip time is Δt and the characteristic impedance is

$$Z_s = \frac{\Delta t}{2C_s} \tag{4.29}$$

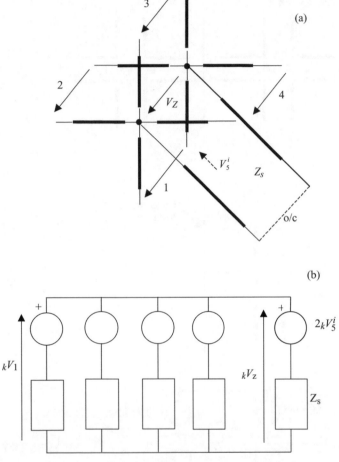

FIGURE 4.11: A shunt node with a stub (a) and its Thevenin equivalent circuit (b)

The stub is inserted as shown in Fig. 4.11(a) so that it presents a "load" to the electric field. The Thevenin equivalent for this configuration is shown in Fig. 4.11(b) and scattering for this node proceeds by calculating $_kV_Z$ and then proceeding as in (4.23).

4.4   SOME PRACTICAL REMARKS

We have seen how to introduce perfect electrical conductors (PEC) in the connection process [see Eq. (4.7)]. This kind of boundary condition (BC) is often referred to as a short-circuit BC. What other kinds of BC do we need?

Consider a problem where there is symmetry across a boundary O–O' as shown in Fig. 4.12(a). Symmetry implies that at every time step voltages reflected into port 4 of node

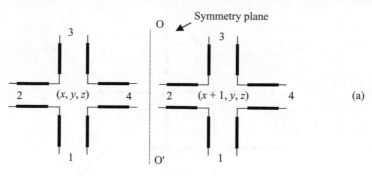

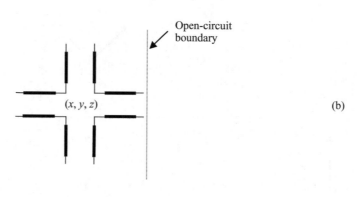

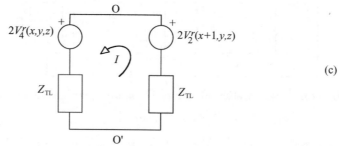

FIGURE 4.12: A symmetry plane OO' (a), its representation by an open-circuit boundary (b), and Thevenin equivalent circuit at the boundary (c)

(x, y, z) and exactly equal to those reflected into port 2 of node $(x + 1, y, z)$, i.e.,

$$_k V_4^r(x, y, z) = {}_k V_2^r(x + 1, y, z) \tag{4.30}$$

Therefore, the incident voltage at $k + 1$ on port 4 of node (x, y, z) is

$$_{k+1} V_4^i(x, y, z) = {}_k V_2^r(x + 1, y, z) = {}_k V_4^r(x, y, z) \tag{4.31}$$

where the second equality in (4.31) stems from (4.30). Looking at the equality of the terms on the extreme left and right of (4.31) we can recognize that as far as node (x, y, z) is concerned the boundary O–O' behaves as an open-circuit boundary (reflection coefficient equal to 1). We need to model only half the problem say the left-hand side as shown in Fig. 4.12(b). Such a symmetry boundary is often referred to as a perfect magnetic conductor (PMC) boundary. The reason is not difficult to see. Let us try to calculate the magnetic field component H_z at the boundary O–O'. Looking left and right we replace what we see with its Thevenin equivalent as shown in Fig. 4.12(c). The current responsible for the z-component of the magnetic field is then,

$$I = \frac{V_2^r(x+1, y, z) - V_4^r(x, y, z)}{Z_{TL}} \qquad (4.32)$$

However, from (4.30) the numerator is always zero therefore on this boundary H_z is always zero—a perfect magnetic conductor. An open-circuit boundary introduces therefore a symmetry plane. If the problem possesses more than one symmetry plane, then several open-circuit BCs may be employed to reduce substantially the size of the computation.

We have avoided so far talking about open-boundary problems. The problem of transmission from a radio transmitter is an open-boundary problem—the field fills the entire space around the antenna of the transmitter as far as the moon and beyond! A computation as we have described cannot possibly populate such a vast space. Inevitably, an open-boundary problem must be "terminated" somehow in the numerical domain so that a manageable computation can be done. Put another way, although there is no physical boundary to the problem we must impose a "numerical" BC. This must be done without generating numerical artefacts that contaminate the solution to the problem inside the domain being modeled. This is a particular problem of all differential equation methods (not just TLM) which require volumetric meshing. This BC must allow any outgoing waves free passage as if the numerical boundary is not present. A lot of work has been done, and continues at present, to design such a perfect absorbing boundary condition (ABC)—it is not an easy matter. It is not appropriate in this introductory text to go into great detail (cf. [14]). However, I will show you how to get a pretty good ABC in TLM with little effort and computational cost. Consider the case where a series mesh meant to model an open-boundary problem is terminated numerically at A–A' as shown in Fig. 4.13. Since I need to put some termination there I have simply introduced an impedance Z to terminate port 4 that abuts to this boundary. What should the value of Z be so that this numerical boundary is as transparent as possible (no reflections)? Well, for start we can calculate the reflection coefficient seen by pulses travelling towards port 4 at the boundary,

$$\Gamma = \frac{Z - Z_{TL}}{Z + Z_{TL}} \qquad (4.33)$$

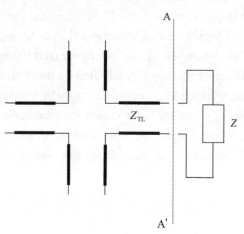

FIGURE 4.13: Termination (matching) of a series node with an impedance Z

You may be tempted to say that Z must be selected to be equal to Z_{TL} so that $\Gamma = 0$. This however is incorrect. We must correctly terminate the medium not the TL! Assuming that we model free space the intrinsic impedance of the medium is $\eta = \eta_0 = \sqrt{\mu_0/\varepsilon_0}$. From (4.17) we have $Z_{TL} = \eta_0/\sqrt{2}$, hence substituting into (4.33) we find that correct termination in this case requires a reflection coefficient given by

$$\Gamma = \frac{\sqrt{2}-1}{\sqrt{2}+1} \qquad (4.34)$$

This is a good (but not perfect) ABC in TLM and is known as a "matched boundary condition." If a wave hits the matched boundary at right angles, absorption is good but less so at grazing angles. In practice, although better ABCs are available, a matched boundary condition is satisfactory provided it is placed some distance away from the region of interest. The effectiveness of the ABC may be established by numerical experimentation (shifting the boundary and checking the results remain essentially unchanged).

Reflecting on the structure of a TLM mesh (a complex network of capacitors and inductors) one can see that it is some kind of low-pass filter. It is therefore to be expected that the performance of the mesh will deteriorate at high frequencies to a point where it becomes unacceptable. Put in a more formal way the mesh is dispersive and anisotropic. In simple cases, such as the 2D meshes considered here, dispersive properties may be assessed analytically. In 3D cases, where the mesh is more complex, dispersion analysis (i.e., assessment of errors with frequency) is a difficult task. More details may be found in [4]. As a rule of thumb, if we have a resolution of ten segments per wavelength velocity errors (percentage deviation of the velocity of propagation from its ideal value) are of the order of a few percent. Using say 20 segments per wavelength increases the accuracy further but at a substantial additional computational cost.

4.5   MODAL VIEW OF TLM

So far I have presented a view of TLM based on an analogy to circuit equations, which to some extent is hand waving and is justified by the physical reasonableness of the basic ideas. This may not satisfy every reader. A more systematic model building from Maxwell's equations may be found in [4] and references therein. What I plan to do in this section is to present yet another way of interpreting the workings of the 2D TLM model that is based on ideas from the modal expansion of solutions to Maxwell's equations. This approach has two attractive features. First, it is very satisfying from the methodological point of view, and second, it leads to a very practical modeling tool as shown in the next section. This section is not essential for understanding and using TLM but it will help you to follow some of the more complex ideas further on. I suggest that first time you read through without getting down to detail to get the basic concepts.

The starting point is Maxwell's equation in cylindrical coordinates [15, 16] where all field components are expressed in terms of H_z and E_z,

$$E_r = -\frac{j}{k_c^2}\left(\beta\frac{\partial E_z}{\partial r} + \frac{\omega\mu}{r}\frac{\partial H_z}{\partial\varphi}\right)$$

$$E_\varphi = \frac{j}{k_c^2}\left(-\frac{\beta}{r}\frac{\partial E_z}{\partial\varphi} + \omega\mu\frac{\partial H_z}{\partial r}\right)$$

$$H_r = \frac{j}{k_c^2}\left(\frac{\omega\varepsilon}{r}\frac{\partial E_z}{\partial\varphi} - \beta\frac{\partial H_z}{\partial r}\right)$$ (4.35)

$$H_\varphi = -\frac{j}{k_c^2}\left(\omega\varepsilon\frac{\partial E_z}{\partial r} + \frac{\beta}{r}\frac{\partial H_z}{\partial\varphi}\right)$$

where $k_c^2 = k^2 - \beta^2$, $k^2 = \omega^2\mu\varepsilon$, and β is the phase constant. Fields are assumed to vary as $\sim \exp j(\omega t \pm \beta z)$. We will illustrate the basic ideas with reference to the shunt node and therefore we assume a TM mode where $H_z = 0$. Equations (4.35) transform the *TM mode* to

$$E_r = -\frac{j\beta}{k_c^2}\frac{\partial E_z}{\partial r}$$

$$E_\varphi = -\frac{j\beta}{k_c^2 r}\frac{\partial E_z}{\partial\varphi}$$

$$H_r = \frac{j\omega\varepsilon}{k_c^2 r}\frac{\partial E_z}{\partial\varphi}$$ (4.36)

$$H_\varphi = -\frac{j\omega\varepsilon}{k_c^2}\frac{\partial E_z}{\partial r}$$

We see that we can recover all field components if E_z is known. This component may be found by solving the following equation obtained directly from Maxwell's equations,

$$\frac{1}{r}\frac{\partial}{\partial r}\left(r\frac{\partial E_z}{\partial r}\right) + \frac{1}{r^2}\frac{\partial^2 E_z}{\partial \varphi^2} + k_c^2 E_z = 0 \tag{4.37}$$

The solution to this equation is obtained by separation of variables (r-dependent and φ-dependent functions) that have the general form,

$$R(r) = C_1 H_n^{(1)}(k_c r) + C_2 H_n^{(2)}(k_c r)$$
$$F(\varphi) = C_3 \cos(n\varphi) + C_4 \sin(n\varphi) \tag{4.38}$$

where C_1–C_4 are constants to be determined by the boundary conditions, $H_n^{(.)}(k_c r)$ are Hankel functions of order n of the first and second kind, respectively. The general solution for the electric field can thus be put in the form,

$$E_z(r, \varphi) = \sum_{n=-\infty}^{+\infty} A_n e^{jn\varphi}\left[H_n^{(1)}(k_c r) + C_n H_n^{(2)}(k_c r)\right] \tag{4.39}$$

where A_n and C_n are constants to be obtained from the boundary conditions and k_c is the same as given in connection with (4.35). We can make the following observations about (4.39):

- The total field is a combination of an *infinite number of modes* (summation over n)

- The Hankel functions are related to the Bessel functions by

$$H_n^{(1)}(k_c r) = J_n(k_c r) + j N_n(k_c r)$$
$$H_n^{(2)}(k_c r) = J_n(k_c r) - j N_n(k_c r) \tag{4.40}$$

  where $J_n(.)$ and $N_n(.)$ are the Bessel functions and Neumann functions, respectively. This looks like the decomposition of exponential functions into trigonometric functions, $e^{\pm j(k_c r)} = \cos(k_c r) \pm j \sin(k_c r)$ therefore, by analogy, the Hankel function of the second kind represents outgoing waves and of the first kind incoming waves.

- Not all terms in (4.40) are present in any particular problem. Depending on the structure and symmetry of the problem some terms are not permissible. For example, $N_n(k_c r)$ has a singularity at $r = 0$ (tends to infinity) and therefore cannot be included as a solution for the field at $r = 0$.

- Expression (4.40) may be substituted into (4.39) to obtain an expression for the field in terms of J_n and N_n,

$$E_z(r, \varphi) = \sum_{n=-\infty}^{+\infty} B_n e^{jn\varphi}\left[J_n(k_c r) + D_n N_n(k_c r)\right] \tag{4.41}$$

where B_n and D_n are the constants to be obtained from the boundary conditions.

The developments described above form the background on which we examine the manner in which a shunt node maps TM fields in free space (no other objects present inside the node). In such a case, we must make sure that the field has a finite value at $r = 0$ hence $N_n(k_c r)$ in (4.41) is not admissible as part of the solution ($D_n = 0$). For convenience, we also normalize the solution at the boundary of the node $\Delta = \Delta\ell/2$ to obtain the form of (4.41) in *free space*,

$$E_z(r, \varphi) = \sum_{n=-\infty}^{+\infty} B_n e^{jn\varphi} \frac{J_n(k_c r)}{J_n(k_c \Delta)} \qquad (4.42)$$

The φ-component of the magnetic field may be obtained from the last expression in (4.36) where we have set $\beta = 0$ (since in the 2D problem there is no variation along z) and $k^2 = \omega^2 \mu_0 \varepsilon_0$ (free space)

$$H_\varphi(r, \varphi) = \frac{1}{j\omega\mu_0} \frac{\partial E_z(r, \varphi)}{\partial r} \qquad (4.43)$$

Expressions (4.42) and (4.43) represent the structure of the 2D TM field in free space that consists of an infinite number of modes. The question we pose is the following: How much of this rich texture of the field can a shunt node represent? From Fig. 4.14 we see that our model is only sampling the field at four points (TL ports). It is therefore reasonable to assert that since we have four degrees of freedom in the model we should be able at best to represent four modes of the infinite expansion in (4.42). We will explore this idea in more detail. Our node models well the EM field if it is so structured as to present the correct admittance relating the electric and magnetic field components. We observe that on the surface of the node we define electric and magnetic field components that are not independent. They are related by an admittance operator Y

$$H = YE \qquad (4.44)$$

In TLM field components are expressed in terms of incident and reflected voltage pulses, i.e.,

$$E = V^i + V^r, \qquad H = Y_L(V^i - V^r) \qquad (4.45)$$

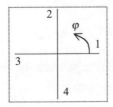

FIGURE 4.14: Schematic of a node in 2D for the study of the modal representation of TLM

where Y_L is a 4×4 diagonal matrix with elements equal to y_L and the admittance of the link lines equal to $\sqrt{\varepsilon_L/\mu_L}$. Thus, $Y_L = y_L I$ where I is the identity matrix. In (4.45) I have suppressed for simplicity the scaling factor $\Delta \ell$. Substituting (4.45) into (4.44) and separating incident voltages from reflected voltages we obtain,

$$\{Y + y_L I\} V^r = \{-Y + y_L I\} V^i \tag{4.46}$$

This expression contains all the information needed to implement the scattering at the node, i.e., to relate the reflected voltages to the incident voltages. The best way to solve this equation for V^r is to solve the eigenvalue problem,

$$Y U_n = \gamma_n U_n \tag{4.47}$$

that is, to find the eigenvalues γ_n and eigenvectors U_n of the admittance operator Y. The eigenvectors are effectively the modes of the field. They form a complete set and are orthogonal (in mathematical parlance Y is self-adjoint), therefore we can represent the incident and reflected voltages as a superposition of modal solutions.

$$V^i = U X^i \tag{4.48}$$

where U is a 4×4 matrix with the eigenvectors of Y as its columns and X^i represents the projection of the incident voltages to the eigenvectors (modes of the field). I write below (4.48) in full to facilitate understanding.

$$\begin{bmatrix} V_1^i \\ V_2^i \\ V_3^i \\ V_4^i \end{bmatrix} = \begin{bmatrix} U_1 \; U_2 \; U_3 \; U_4 \end{bmatrix} \begin{bmatrix} X_1^i \\ X_2^i \\ X_3^i \\ X_4^i \end{bmatrix} \tag{4.49}$$

The column vector on the left has as its components the four incident port voltages. Each column of the matrix (four rows) represents a field mode (eigenvector of the admittance operator) and the elements of the column vector on the right are the projections of each incident voltage onto each mode (how much of a particular mode is present in each incident voltage). Similar expressions are obtained for the projection of the reflected voltages onto the four modes.

$$V^r = U X^r \tag{4.50}$$

We may now substitute the modal expansions of the incident and reflected voltages [Eqs. (4.48) and (4.50)] into the scattering equation (4.46),

$$y_L I U X^r + Y U X^r = y_L I U X^i - Y U X^i \tag{4.51}$$

I will evaluate in detail the two terms on the left-hand side,

$$y_L IUX^r = y_L \begin{bmatrix} U_1 & U_2 & U_3 & U_4 \end{bmatrix} \begin{bmatrix} X_1^r \\ X_2^r \\ X_3^r \\ X_4^r \end{bmatrix}$$

$$= \begin{bmatrix} y_L U_1 X_1^r & y_L U_2 X_2^r & y_L U_3 X_3^r & y_L U_4 X_4^r \end{bmatrix} \rightarrow y_L U_n X_n^r \qquad (4.52)$$

and similarly,

$$YUX^r = Y \begin{bmatrix} U_1 & U_2 & U_3 & U_4 \end{bmatrix} \begin{bmatrix} X_1^r \\ X_2^r \\ X_3^r \\ X_4^r \end{bmatrix}$$

$$= \begin{bmatrix} YU_1 & YU_2 & YU_3 & YU_4 \end{bmatrix} \begin{bmatrix} X_1^r \\ X_2^r \\ X_3^r \\ X_4^r \end{bmatrix}$$

$$= \begin{bmatrix} \gamma_1 U_1 X_1^r & \gamma_2 U_2 X_2^r & \gamma_3 U_3 X_3^r & \gamma_4 U_4 X_4^r \end{bmatrix} \rightarrow \gamma_n U_n X_n^r \qquad (4.53)$$

The right-hand side of (4.51) is evaluated in the same way to give

$$\begin{bmatrix} (y_L + \gamma_1) X_1^r U_1 & (y_L + \gamma_2) X_2^r U_2 & (y_L + \gamma_3) X_3^r U_3 & (y_L + \gamma_4) X_4^r U_4 \end{bmatrix} =$$

$$\begin{bmatrix} (y_L - \gamma_1) X_1^i U_1 & (y_L - \gamma_2) X_2^i U_2 & (y_L - \gamma_3) X_3^i U_3 & (y_L - \gamma_4) X_4^i U_4 \end{bmatrix} \qquad (4.54)$$

Equating the coefficient of U_n on the left- and right-hand side we get

$$X_n^r = \frac{y_L - \gamma_n}{y_L + \gamma_n} X_n^i, \qquad n = 1, 2, 3, 4 \qquad (4.55)$$

Equation (4.55) shows how the reflected modal components of the voltage may be obtained from the incident modal components. It is an elegant expression illustrating how field scattering takes place at the modal level. The physical interpretation of the eigenvalues is that they represent the admittance seen by each mode propagating on the TLs comprising the TLM node, as shown schematically in Fig. 4.15. From TL theory and this figure we see that

$$X_n^r = \frac{\dfrac{1}{\gamma_n} - \dfrac{1}{y_L}}{\dfrac{1}{\gamma_n} + \dfrac{1}{y_L}} X_n^i = \frac{y_L - \gamma_n}{y_L + \gamma_n} X_n^i \text{ as expected from } (4.55)$$

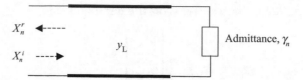

FIGURE 4.15: The eigenvalues γ_n as the admittance seen by mode n

The question we must address now is how to obtain the eigenvalues. The simplest way is to use the known modal structure of the field [Eqs. (4.42) and (4.43)] to obtain the admittance of each mode and hence the required eigenvalues. We see that because the origin of φ in these expressions is arbitrary we can replace the exponential term by $\cos(n\varphi)$, without loss of generality, to obtain the following expressions for the modal field components.

$$E_{zn}(r, \varphi) = B_n \frac{J_n(k_c r)}{J_n(k_c \Delta)} \cos(n\varphi)$$

$$H_{\varphi n}(r, \varphi) = B_n \frac{1}{j\omega\mu_0} \frac{\partial J_n(k_c r)/\partial r}{J_n(k_c \Delta)} \cos(n\varphi)$$

(4.56)

We select the four lowest order modes since our model has only four degrees of freedom to obtain the following variations in φ.

$$n = 0 \rightarrow \cos(n\varphi) = 1$$
$$n = 1 \rightarrow \cos\varphi$$
$$n = -1 \rightarrow \sin\varphi$$
$$n = 2 \rightarrow \cos(2\varphi)$$

Referring to Fig. 4.14 we see that the φ-variation of the four modes is given as shown in Table 4.1 where we have normalized each mode $\left(\sqrt{V_{1n}^2 + V_{2n}^2 + V_{3n}^2 + V_{4n}^2} = 1\right)$.

TABLE 4.1: Normalised amplitude of φ-variation of the first four lowest order modes.

	V_1	V_2	V_3	V_4
$n = 0$	$1/2$	$1/2$	$1/2$	$1/2$
$n = 0$	$1/\sqrt{2}$	0	$-1/\sqrt{2}$	0
$n = -1$	0	$1/\sqrt{2}$	0	$-1\sqrt{2}$
$n = 2$	$1/2$	$-1/2$	$1/2$	$-1/2$

The four modal electric field components are thus

$$E_{z1} = \frac{1}{2}B_0 \left\{ \frac{J_0(k_c r)}{J_0(k_c \Delta)} \right\}$$

$$E_{z2} = \frac{1}{\sqrt{2}}B_1 \left\{ \frac{J_1(k_c r)}{J_1(k_c \Delta)} \right\} \cos(\varphi)$$

$$E_{z3} = \frac{1}{\sqrt{2}}B_{-1} \left\{ \frac{J_1(k_c r)}{J_1(k_c \Delta)} \right\} \sin(\varphi)$$

$$E_{z4} = \frac{1}{2}B_2 \left\{ \frac{J_2(k_c r)}{J_2(k_c \Delta)} \right\} \cos(2\varphi)$$

(4.57)

Similar expressions may also be obtained for the magnetic field using the second equation in (4.56). The strength of the incident component of each mode is obtained from the incident port voltages from the expression,

$$
\begin{bmatrix} X_0^i \\ X_1^i \\ X_{-1}^i \\ X_2^i \end{bmatrix}
=
\begin{bmatrix}
\frac{1}{2} & \frac{1}{2} & \frac{1}{2} & \frac{1}{2} \\
\frac{1}{\sqrt{2}} & 0 & -\frac{1}{\sqrt{2}} & 0 \\
0 & \frac{1}{\sqrt{2}} & 0 & -\frac{1}{\sqrt{2}} \\
\frac{1}{2} & -\frac{1}{2} & \frac{1}{2} & -\frac{1}{2}
\end{bmatrix}
\begin{bmatrix} V_1^i \\ V_2^i \\ V_3^i \\ V_4^i \end{bmatrix}
$$

(4.58)

Equation (4.58) may be solved for the incident port voltages,

$$
\begin{bmatrix} V_1^i \\ V_2^i \\ V_3^i \\ V_4^i \end{bmatrix}
=
\begin{bmatrix}
\frac{1}{2} & \frac{1}{\sqrt{2}} & 0 & \frac{1}{2} \\
\frac{1}{2} & 0 & \frac{1}{\sqrt{2}} & -\frac{1}{2} \\
\frac{1}{2} & -\frac{1}{\sqrt{2}} & 0 & \frac{1}{2} \\
\frac{1}{2} & 0 & -\frac{1}{\sqrt{2}} & -\frac{1}{2}
\end{bmatrix}
\begin{bmatrix} X_0^i \\ X_1^i \\ X_{-1}^i \\ X_2^i \end{bmatrix}
$$

(4.59)

Comparing this expression with (4.49) we see that the matrix above is the matrix of eigenvectors U. The matrix in (4.58) is the transpose of U. Thus the last two expressions can be put in a compact form,

$$
\begin{aligned}
X_n^i &= U^T V_n^i \\
V_n^i &= U X_n^i
\end{aligned}
$$

(4.60)

These expressions allow us to transform port incident voltages to modal incident voltages and vice versa. Similar expressions apply for the reflected voltages. The strategy is straightforward: at each time step we decompose the incident port voltages into incident modal components using the first equation in (4.60); the reflected modal components are then obtained using (4.55) provided we know the eigenvalues; the reflected modal components are combined using the second expression in (4.60) (simply replace superscript i by r) to obtain the reflected nodal voltages; these are passed on to the nearest neighbors (connection) and so on! I still need to calculate the eigenvalues. This is straightforward now. I will show this in detail first for the lowest order mode $n = 0$.

From (4.56) the ratio of H to E evaluated at the edge of the node ($r = \Delta$) is

$$Y_n = \frac{H_\varphi(r, \varphi)}{E_z(r, \varphi)}\Big|_{r=\Delta} = \frac{1}{j\omega\mu_0 J_n(k_c \Delta)} \frac{\partial J_n(k_c r)}{\partial r}\Big|_{r=\Delta} \qquad (4.61)$$

At this point we must make a reasonable approximation. The argument of the Bessel function $(k_c r)$ where r is of the order of Δ is very small. Remember that $k_c \Delta = (2\pi /\lambda)\, \Delta\ell/2 \sim \Delta\ell/\lambda \ll 1$ if the model has been correctly constructed. Therefore, we can approximate the Bessel functions by their small argument approximations (see Appendix 2).

$$J_0(k_c r) \simeq 1 - \frac{k_c^2 r^2}{4}$$

$$J_1(k_c r) \simeq \frac{k_c r}{4} \qquad (4.62)$$

$$J_2(k_c r) \simeq \frac{k_c^2 r^2}{8}$$

Hence, from (4.61) we obtain for $n = 0$,

$$Y_0 = \frac{-(k_c^2/4)2\Delta}{j\omega\mu_0 \left[1 - (k_c \Delta/2)^2\right]} \simeq \frac{j\omega\varepsilon_0 \Delta}{2} \qquad (4.63)$$

where we have neglected the second term in the denominator compared to one. This is the admittance seen by this mode at the edge of the node. For $n \neq 0$ we get in a similar manner,

$$Y_n \simeq \frac{n}{j\omega\mu_0\Delta} \qquad (4.64)$$

We are now set to bring the whole thing together. We can clarify even further the nature of the scattering of the four modes in Eq. (4.55). The eigenvalues (admittances presented to each mode) are imaginary and therefore, as expected, in a lossless situation scattering will be simply a matter of a phase shift. In TLM this can be achieved by launching the modal component into TL segments, i.e.,

$$X_n^r = \pm e^{-j\sigma_n} X_n^i \tag{4.65}$$

where σ_n is the phase shift along the line. For an open circuit segment the plus sign applies and we see that the reflection coefficient is

$$+e^{-j\sigma_n} = \frac{e^{-j\sigma_n/2}}{e^{j\sigma_n/2}} = \frac{\cos(\sigma_n/2) - j\sin(\sigma_n/2)}{\cos(\sigma_n/2) + j\sin(\sigma_n/2)}$$

$$= \frac{1 - j\tan(\sigma_n/2)}{1 + j\tan(\sigma_n/2)} \simeq \frac{1 - j(\sigma_n/2)}{1 + j(\sigma_n/2)} \tag{4.66}$$

For a short circuit segment the minus sign applies and the reflection coefficient is then,

$$-e^{-j\sigma_n} = -\frac{1 - j\tan(\sigma_n/2)}{1 + j\tan(\sigma_n/2)} \simeq -\frac{1 - j(\sigma_n/2)}{1 + j(\sigma_n/2)} \tag{4.67}$$

For the $n = 0$ mode the desired reflection coefficient is

$$\frac{y_L - \gamma_0}{y_L + \gamma_0} = \frac{y_L - (j\omega\varepsilon_0\Delta/2)}{y_L + (j\omega\varepsilon_0\Delta/2)} = \frac{1 - (j\omega\varepsilon_0\Delta/2y_L)}{1 + (j\omega\varepsilon_0\Delta/2y_L)}$$

Comparing this expression with (4.66) for an open-circuit line we see that it requires a shift σ_0 such that

$$\frac{j\omega\varepsilon_0\Delta}{2y_L} = j\frac{\sigma_0}{2} \rightarrow \sigma_0 \simeq \frac{\omega\varepsilon_0\Delta}{y_L} \tag{4.68}$$

For the case of $n \neq 0$, using (4.64) we obtain the reflection coefficient,

$$\frac{y_L - \gamma_n}{y_L + \gamma_n} = \frac{y_L - \dfrac{n}{j\omega\mu_0\Delta}}{y_L + \dfrac{n}{j\omega\mu_0\Delta}} = -\frac{1 - \dfrac{j\omega\mu_0\Delta y_L}{n}}{1 + \dfrac{j\omega\mu_0\Delta y_L}{n}} \tag{4.69}$$

Comparing (4.69) with (4.67) we can see that this reflection coefficient may be implemented as a short-circuit line with a phase shift,

$$\sigma_n \simeq \frac{2y_L\omega\mu_0\Delta}{n} \tag{4.70}$$

In TLM this can be achieved by launching the modal components into TL segments. The complete scheme is set out in Fig. 4.16. Taking the open-circuit segment as an example,

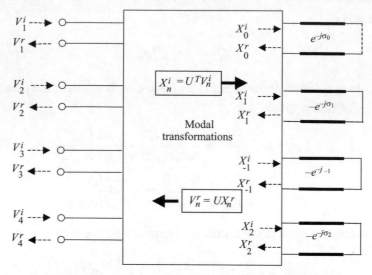

FIGURE 4.16: Transformation from total incident nodal voltages to modal incident voltages, and the reverse process after reflection

length Δ and round trip time Δt, we see that the phase shift conditions are

$$e^{-j\sigma_0} = e^{-j\omega \frac{\varepsilon_0 \Delta}{y_L}} = e^{-j\omega \Delta t}$$

$$\rightarrow \Delta t = \frac{\varepsilon_0 \Delta}{y_L}$$

Choosing $y_L = y_0/\sqrt{2}$ we obtain as the velocity of propagation on this line,

$$u = \frac{2\Delta}{(\Delta t)} = \frac{2y_L}{\varepsilon_0} = \frac{2y_0}{\sqrt{2}\varepsilon_0} = \sqrt{2}\frac{1}{\sqrt{\varepsilon_0\mu_0}} = \sqrt{2}c \qquad (4.71)$$

Similarly, for $n \neq 0$ we get from the short-circuit segments,

$$\Delta t = \frac{2y_L\mu_0\Delta}{n}$$

and hence a velocity of propagation,

$$u = \frac{2\Delta}{(\Delta t)} = n\sqrt{2}c \qquad (4.72)$$

We see from (4.71) and (4.72) that the velocity of propagation for modes $n = 0, 1$ is the same. The exception is mode $n = 2$ for which the velocity is twice as large. Without significant loss in accuracy we choose the same velocity even for the mode $n = 2$ as this simplifies calculations. The higher the order of a mode the greater the spatial frequency (more structure) and hence these modes suffer significant dispersion. Our approximation therefore will not have a major impact. We note that the velocity of propagation (4.71) and the line admittance y_L are consistent with the choices made in (4.27) and (4.28) in connection with the conventional derivation of the 2D TLM node. I repeat once again the procedure outlined in Fig. 4.16. The total incident voltage at the four ports of a node are decomposed into their four modal components [first expression in (4.60)]; modal components are scattered (phase shifted) as already indicated; the reflected (scattered) modal components are then combined to obtain the total reflected components [second expression in (4.60)]; the total reflected components are passed on to the nearest neighbors (connection); the process repeats as long as required.

The process above is essentially presented as a signal-processing task. We may however, if we wish so, produce an equivalent circuit that automatically implements the decomposition into modes and vice versa. This is another way of viewing TLM—a circuit modal analyzer and scatterer! I will confirm that this is the case by demonstrating that the scattering matrix S for the shunt node is obtained directly from the modal picture outlined above [17].

We start with the equations relating total port voltages and modal component [analogous to (4.60)],

$$X_n = U^T V_n$$
$$V_n = U X_n \tag{4.73}$$

The relationship between the port voltage and current vectors is analogous to (4.61), i.e.,

$$I_n = U\gamma X_n \tag{4.74}$$

where γ is a diagonal matrix with the eigenvalues of the admittance matrix given by Eqs. (4.63) and (4.64) as its components. Combining the last two equations we obtain

$$I_n = \{U\gamma U^T\} V_n = Y V_n \tag{4.75}$$

But in TLM we can obtain the voltages and currents in terms of the incident and reflected pulses,

$$V_n = V_n^i + V_n^r \tag{4.76}$$
$$I_n = y_L I (V_n^i - V_n^r) \tag{4.77}$$

where y_L is a scalar representing the line admittance and I is the identity matrix. Equation (4.75) may be substituted into (4.77) to give

$$U\gamma U^T \left(V_n^i + V_n^r \right) = y_L I \left(V_n^i - V_n^r \right) \tag{4.78}$$

This equation relates incident and reflected voltages and hence contains all the information about the scattering matrix. Grouping incident and reflected voltage terms separately we get

$$\left(U\gamma U^{T} - y_{L}I\right) V_{n}^{i} + \left(U\gamma U^{T} + y_{L}I\right) V_{n}^{r} = 0 \qquad (4.79)$$
$$U\left(\gamma - y_{L}I\right) U^{T} V_{n}^{i} + U\left(\gamma + y_{L}I\right) U^{T} V_{n}^{r} = 0 \qquad (4.80)$$

Hence, the reflected voltage vector is

$$
\begin{aligned}
V_{n}^{r} &= -(U^{T})^{-1} \left(\gamma - y_{L}I\right)^{-1} U^{-1} U (\gamma + y_{L}I) U^{T} V_{n}^{i} \\
&= U(y_{L}I + \gamma)^{-1}(y_{L}I - \gamma) U^{T} V_{n}^{i} \\
&= UAU^{T} V_{n}^{i} = SV_{n}^{i} \qquad (4.81)
\end{aligned}
$$

where S is the required scattering matrix.

The matrix A is diagonal and can be obtained directly by recognizing that

$$
y_{L}I =
\begin{bmatrix}
y_{L} & & & \\
& y_{L} & & \\
& & y_{L} & \\
& & & y_{L}
\end{bmatrix}, \qquad
\gamma =
\begin{bmatrix}
\gamma_{0} & & & \\
& \gamma_{1} & & \\
& & \gamma_{2} & \\
& & & \gamma_{3}
\end{bmatrix}
$$

where the eigenvalues are given by (4.63) and (4.64). Carrying out the multiplications in (4.81) we obtain

$$
A =
\begin{bmatrix}
a_{1} & & & \\
& a_{2} & & \\
& & a_{3} & \\
& & & a_{4}
\end{bmatrix}
$$

where

$$
a_{1} = \frac{y_{L} - \dfrac{j\omega\varepsilon\Delta}{2}}{y_{L} + \dfrac{j\omega\varepsilon\Delta}{2}} \qquad
a_{2} = \frac{y_{L} - \dfrac{1}{j\omega\mu\Delta}}{y_{L} + \dfrac{1}{j\omega\mu\Delta}}
$$

$$(4.82)$$

$$
a_{3} = \frac{y_{L} - \dfrac{1}{j\omega\mu\Delta}}{y_{L} + \dfrac{1}{j\omega\mu\Delta}} \qquad
a_{4} = \frac{y_{L} - \dfrac{2}{j\omega\mu\Delta}}{y_{L} + \dfrac{2}{j\omega\mu\Delta}}
$$

The scattering matrix can then be obtained by carrying out the operations on the RHS of

$$S = UAU^T \tag{4.83}$$

After some algebra we obtain

$$S = \begin{bmatrix} \frac{1}{4}(a_1 + 2a_2 + a_4) & \frac{1}{4}(a_1 - a_4) & \frac{1}{4}(a_1 - 2a_2 + a_4) & \frac{1}{4}(a_1 - a_4) \\ \frac{1}{4}(a_1 - a_4) & \frac{1}{4}(a_1 + 2a_3 + a_4) & \frac{1}{4}(a_1 - a_4) & \frac{1}{4}(a_1 - 2a_3 + a_4) \\ \frac{1}{4}(a_1 - 2a_2 + a_4) & \frac{1}{4}(a_1 - a_4) & \frac{1}{4}(a_1 + 2a_2 + a_4) & \frac{1}{4}(a_1 - a_4) \\ \frac{1}{4}(a_1 - a_4) & \frac{1}{4}(a_1 - 2a_3 + a_4) & \frac{1}{4}(a_1 - a_4) & \frac{1}{4}(a_1 + 2a_3 + a_4) \end{bmatrix} \tag{4.84}$$

The elements of the scattering matrix in (4.84) are in the frequency-domain. To map into the time-domain we apply the bilinear transformations [18]

$$j\omega \to \frac{2}{\Delta t} \frac{1 - z^{-1}}{1 + z^{-1}} \tag{4.85}$$

where z^{-1} represents a delay of Δt. Applying (4.85) to each quantity at a time we obtain

$$a_1 = \frac{y_L - \dfrac{\varepsilon \Delta}{\Delta t} \dfrac{1 - z^{-1}}{1 + z^{-1}}}{y_L + \dfrac{\varepsilon \Delta}{\Delta t} \dfrac{1 - z^{-1}}{1 + z^{-1}}} = \frac{\left(y_L - \dfrac{\varepsilon \Delta}{\Delta t}\right) + z^{-1}\left(y_L + \dfrac{\varepsilon \Delta}{\Delta t}\right)}{\left(y_L + \dfrac{\varepsilon \Delta}{\Delta t}\right) + z^{-1}\left(y_L - \dfrac{\varepsilon \Delta}{\Delta t}\right)} \tag{4.86}$$

If in (4.86) we choose

$$y_L = \frac{\varepsilon \Delta}{\Delta t} \tag{4.87}$$

then the first diagonal element of A represents simply a time delay Δt

$$a_1 = z^{-1}$$

In a similar fashion, we obtain

$$a_2 = \frac{\left(y_L - \dfrac{\Delta t}{2\mu \Delta}\right) - z^{-1}\left(y_L + \dfrac{\Delta t}{2\mu \Delta}\right)}{\left(y_L + \dfrac{\Delta t}{2\mu \Delta}\right) - z^{-1}\left(y_L - \dfrac{\Delta t}{2\mu \Delta}\right)} \tag{4.88}$$

If we choose in (4.88) the line admittance as

$$y_\text{L} = \frac{\Delta t}{2\mu\Delta} \tag{4.89}$$

then we get, $a_2 = a_3 = -z^{-1}$, i.e., a one time step delay but with opposite polarity. We note that the choices in (4.87) and (4.89) imply that we must enforce the equality

$$y_\text{L} = \frac{\Delta t}{2\mu\Delta} = \frac{\varepsilon\Delta}{\Delta t} \tag{4.90}$$

Hence, the velocity of propagation on the TLs is

$$u_\text{TL} \equiv \frac{2\Delta}{\Delta t} = \sqrt{2}\frac{1}{\sqrt{\mu\varepsilon}} = \sqrt{2}c \tag{4.91}$$

and the line admittance is

$$y_\text{L} = \frac{\Delta t}{2\mu\Delta} = \frac{1}{2\mu}\frac{2}{\sqrt{2}c} = \frac{1}{\sqrt{2}}\sqrt{\frac{\varepsilon}{\mu}} \tag{4.92}$$

Equations (4.91) and (4.92) are in agreement with (4.27) and (4.28) obtained in the conventional derivation of the 2D TLM node. To return back to the derivation of the scattering matrix we note that for mode $n = 2$ we are not going to get a simple delay [factor of 2 in (4.64)]. To simplify the algorithm we assume that the impedance for this mode is the same as for mode $n = 1$, to obtain $a_4 \simeq -z^{-1}$. This is an approximation in line with the discussion following Eq. (4.72). We are now in a position to evaluate all the elements of the scattering matrix (4.84). I will evaluate as an example elements S_{11} and S_{21}.

$$S_{11} = \frac{1}{4}(a_1 + 2a_2 + a_4) = z^{-1}\frac{1}{4}(1 - 2 - 1) = -\frac{1}{2}z^{-1}$$

$$S_{12} = \frac{1}{4}(a_1 - a_4) = \frac{1}{2}z^{-1}$$

Similar expressions are obtained for the remaining elements to give the scattering matrix

$$S = \frac{1}{2}\begin{bmatrix} -1 & 1 & 1 & 1 \\ 1 & -1 & 1 & 1 \\ 1 & 1 & -1 & 1 \\ 1 & 1 & 1 & -1 \end{bmatrix} \tag{4.93}$$

Fortunately, (4.93) is the same expression as obtained from the conventional TLM formulations shown in (4.24)! It all makes sense, reaching (4.93) through the modal expansion of the fields is an elegant methodological derivation revealing the "sinews" of the EM field and of the TLM model.

Finally, it is easy to show that in our 2D shunt TLM node each mode sees an open-circuit or short-circuit TL segment as implied by the discussion following Eq. (4.65). From Fig. 4.14, assuming that only mode $n = 0$ incident voltage pulses are applied at the four ports, we obtain after deriving the Thevenin equivalent circuit and applying the parallel generator theorem that the total voltage is

$$V = \frac{y_L 2V_1^i + y_L 2V_2^i + y_L 2V_3^i + y_L 2V_4^i}{4y_L} \qquad (4.94)$$

where from the first row of Table 4.1, $V_1^i = V_2^i = V_3^i = V_4^i = 1/2$. Substituting in (4.94) we obtain that $V = 1$. Hence the reflected voltage on each of the four ports is the same and equal to the total voltage minus the incident voltage, i.e., 1/2. This means that this mode sees effectively an open circuit as the reflected voltage is equal to the incident voltage. Applying (4.94) and following the same procedure for the remaining three modes (the rows of Table 4.1) we show in each case that the total voltage is equal to zero and therefore from (3.21) that the reflected voltage is equal to the incident voltage but with the opposite polarity, i.e., these modes see effectively a short circuit as already indicated.

To summarize this section, I have shown that mapping the four incident port voltages to four modes in the cylindrical expansion of the field allows a systematic procedure for calculating the reflected modal components and thus for obtaining the reflected port voltages. Through this procedure we recover exactly the same TLM parameters and algorithm as from the standard derivation outlined in Sections 4.2 and 4.3. This elegant derivation would have been only just that—a theoretical oddity—if it did not give us an insight into how we may address some complex practical problems. This you will see in the following section.

4.6   EMBEDDING A THIN WIRE IN A 2D TLM MESH

I have shown how a 2D mesh can be constructed to model EM fields in an inhomogeneous space and in the presence of conducting boundaries. However, most practical engineering problems make further demands on the modeler. The most obvious one is the modeling of wires. How can we model a wire in a 2D mesh? I will deal first with the case where the wire radius "a" is larger than the mesh resolution $\Delta \ell$. The situation is as depicted in Fig. 4.17(a). Clearly, the cross-section of the wire can be mapped onto several nodes by inserting short circuits at the node ports nearest to the wire perimeter. This is shown in Fig. 4.147(b) where we see that the smooth perimeter of the wire is approximated by stair-cased boundary. All we need to do is to tell the algorithm to recognize during the connection phase the short circuits present at the nodes indicated. We see that we have a "grainy" approximation to the smooth outline of the wire. This is referred to as "stair-casing error" and the finer the resolution of the mesh the smaller it becomes. As the mesh resolution gets coarser the approximation of the wire cross-section

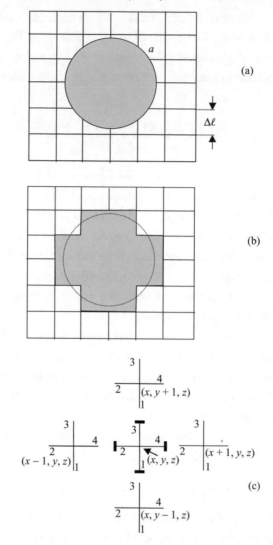

FIGURE 4.17: Mapping the round wire cross-section on a Cartesian mesh (a), stair-case approximation (b), and representing a wire cross-section by a single node with short circuits at its extremities (c)

becomes poorer. The case when $a = \Delta\ell/2$ is depicted in Fig. 4.17(c). All we need to do is to force the nodes around node (x, y, z) to recognize the short circuits shown at the connection phase, i.e.,

$$_{k+1}V_1^i(x, y+1, z) = -_kV_1^r(x, y+1, z)$$
$$_{k+1}V_2^i(x+1, y, z) = -_kV_2^r(x+1, y, z)$$
$$_{k+1}V_3^i(x, y-1, z) = -_kV_3^r(x, y-1, z)$$
$$_{k+1}V_4^i(x-1, y, z) = -_kV_4^r(x-1, y, z)$$

$$(4.95)$$

We must recognize, however, that the approximation to the wire radius obtained in this way is only a rough one. As a rule of thumb, depending on the application, we need a cluster of about 4×4 nodes to describe with some accuracy the wire cross-section.

However, it gets worse when $a < \Delta\ell/2$. What are we going to do then? The obvious option of decreasing $\Delta\ell$ to make it smaller than the wire radius is available to us but at a substantial computational cost which in most cases is prohibitive. To illustrate this point, consider a problem in free space where the highest frequency of interest is 1 GHz. At this frequency the wavelength is 30 cm so an acceptable choice of space discretization is one tenth, i.e., 3 cm. If however we need to model a wire of diameter equal to 3 mm we must increase the spatial resolution in each dimension by at least a factor of ten to achieve a rough description as shown in Fig. 4.17(c). In 2D this represents an increase in storage by a factor of 100 and in 3D by a factor of 1000! The storage costs scale up very rapidly and becomes unrealistic very quickly. Similar adverse scaling takes place as regards the required run time. In order that we advance the computation for the same total period of time we reason that the reduction of $\Delta\ell$ by a factor of ten means that Δt is also reduced by the same factor therefore for the same total problem observation time we will need ten times more time steps. In 2D we need 100 times more calculations per time step and ten times more time steps, i.e., run time scales up by a factor of 1000. Matters look even more unfavorable in 3D!

The purpose of this discussion is to convince you that trying to accommodate fine features such as thin wires by increasing mesh resolution (the brute force way) leads rapidly to unmanageable computations. This is an example of what is referred to as a *multiscale* problem. We have in the same problem features that are electrically large (where resolution of the order $\ell/10$ is adequate) and electrically small (where resolution of $\ell/10$ is insufficient). We need new ways of thinking to deal with multiscale problems if we are going to use computational resources in an intelligent and efficient way! The purpose of this section is to introduce ideas and techniques that can be used for this purpose. The real challenge is to solve multiscale problems in 3D, but we can illustrate better the basic concept in 2D with the minimum of mathematical complexity. A review of multiscale techniques at a more advanced level than in this text may be found in [19].

Two possible approaches are available to us. First, *mesh distortion* whereby the mesh is refined locally (only around the fine feature) thus saving in computational overhead in other areas of the problem where a fine mesh is not required. This is shown in Fig. 4.18(a). This appears to be an attractive option but there are problems. Consider in this figure the interface between the fine and coarse mesh regions. It is obvious that one-to-one correspondence at this interface has been lost. A "coarse" node has several "fine" node neighbors. Hence at the connection phase some means must be worked out of sharing and combining pulses. Similarly, synchronism has been lost as the time step in the fine mesh is different compared to the coarse mesh. Some form of space and time "filtering" is required at the interface between regions of

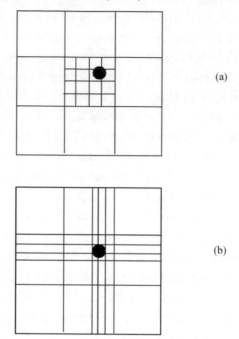

FIGURE 4.18: A small diameter wire represented in a multigrid mesh (a) and in a hybrid or variable mesh (b)

different resolution. Schemes are available to do this but they are difficult to operate and may lead to losses or instability. Such schemes are referred to as *multigrid* schemes and they are the subjects of continuing research interest. Mesh distortion may be implemented as shown in Fig. 4.18(b). Here, a fine mesh region is generated around the fine feature but it is not so well localized as in Fig. 4.18(a). This mesh is known as a hybrid or variable mesh and has the obvious advantage that one-to-one spatial correspondence has been reestablished. Synchronism may also be achieved by a judicious use of stubs but at a reduced time step. We see here that the shape of each node is a general cuboid and this allows us to fit better to various features present in the problem. This distortion is not achieved without a price. The use of stubs results in a higher dispersion compared to a regular mesh (one with cubes as the basic cell) and care must be taken to control errors. More details for the hybrid mesh may be found in [4] and for multigrid and other multiscale techniques in [19].

The second option is to keep a regular coarse mesh but to *embed local solutions* into the mesh that represent the local EM behavior of the fine feature. We focus on this approach here. It follows the ideas developed in Section 4.5. To illustrate more clearly this discussion I address the specific problem of a thin wire of radius *a* placed centrally at a 2D node as shown in Fig. 4.19. This region of space containing the wire interacts with the rest of the problem through

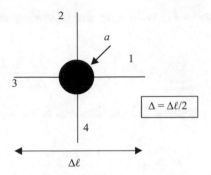

FIGURE 4.19: Representing a fine wire (radius $a \leq \Delta$) in a 2D node

the four port voltages. If we are able to present to incident voltages on these ports an admittance, which correctly represents the relationship between H and E in this region of space (including the impact of the wire), then we can obtain the reflected voltages and implement connection to the rest of the problem. The presence of the thin wire makes its impact through the modification of the admittances compared to the case of free space. The best way of implementing this is to follow the same procedure as for the free space case (Section 4.5) by analyzing port voltages into modes, reflect each mode using the appropriate admittance and then combine the reflected modes to obtain the reflected mode voltages. The formal steps are the same as for the free space case the only differences are in the details–the nature of each mode and its admittance. Note, that in this way, the spatial resolution of the mesh ($\Delta\ell$) is independent of the size of the wire (radius a). We can get away with quite a coarse mesh ($a \ll \Delta\ell$) and still get an accurate wire description. This substantially reduces computational demands. It is for these high stakes that we are playing in this section.

Our starting point is Eq. (4.41) which shows the field expansion in cylindrical coordinates. In the case of free space the recognition of the singularity at $r = 0$ meant that D_n had to be zero thus giving Eq. (4.42). In the present case of a wire at the center of the node the constant D_n must be chosen so that the tangential electric field on the surface of the wire $r = a$ is zero, i.e.,

$$E_z(a, \varphi) = \sum_{n=-\infty}^{+\infty} B_n e^{jn\varphi} \left[J_n(k_c a) + D_n N_n(k_c a) \right] = 0 \qquad (4.96)$$

Hence,

$$D_n = -\frac{J_n(k_c a)}{N_n(k_c a)} \qquad (4.97)$$

Substituting (4.97) into (4.41) we obtain the following expression for the electric field around a wire of radius r

$$E_z(r, \varphi) = \sum_{n=-\infty}^{+\infty} B_n e^{jn\varphi} \left[J_n(k_c r) - \frac{J_n(k_c a)}{N_n(k_c a)} N_n(k_c r) \right] \qquad (4.98)$$

The expression for the magnetic field is obtained from (4.43) which is repeated here for convenience.

$$H_\varphi(r, \varphi) = \frac{1}{j\omega\mu_0} \frac{\partial E_z(r, \varphi)}{\partial r} \qquad (4.99)$$

The calculation now proceeds as for free space by identifying the modes and corresponding admittances. Starting with mode $n = 0$ we have by dividing (4.98) by (4.99)

$$\frac{E_z}{H_\varphi}\Big|_{n=0} = j\omega\mu_0 \frac{J_0(k_c r)N_0(k_c a) - J_0(k_c a)N_0(k_c r)}{N_0(k_c a)\frac{d}{dr}J_0(k_c r) - J_0(k_c a)\frac{d}{dr}N_0(k_c r)} \qquad (4.100)$$

We evaluate the numerator and denominator separately using small argument expansions for the Bessel functions (Appendix 2).

$$\text{num} = \left[1 - \frac{(k_c r)^2}{4} \right] \frac{2}{\pi} \left[\ln\left(\frac{k_c a}{2}\right) + \gamma \right] - \left[1 - \frac{(k_c a)^2}{4} \right] \frac{2}{\pi} \left[\ln\left(\frac{k_c r}{2}\right) + \gamma \right]$$

$$\simeq \frac{2}{\pi} \left[\ln\left(\frac{k_c a}{2}\right) + \gamma \right] - \frac{2}{\pi} \left[\ln\left(\frac{k_c r}{2}\right) + \gamma \right] \simeq \frac{2}{\pi} \ln\left(\frac{a}{r}\right)$$

where we have neglected terms $(k_c r)^2$, $(k_c a)^2$ relative to 1. Similarly, for the denominator we obtain

$$\text{denom} \simeq \frac{2}{\pi} \left[\ln\left(\frac{k_c a}{2}\right) + \gamma \right] \left(-\frac{k_c^2 r}{2} \right) - \frac{2}{\pi} \frac{1}{r}$$

Hence, the impedance for mode $n = 0$ is obtained from (4.100) evaluated at the node's edge $r = \Delta$.

$$\frac{E_{z0}}{H_{z0}}\Big|_{r=\Delta} = j\omega\mu_0\Delta \frac{\ln\left(\frac{a}{\Delta}\right)}{-\frac{(k_c \Delta)^2}{2} \left[\ln\left(\frac{k_c a}{2}\right) + \gamma \right] - 1} \simeq j\omega\mu_0\Delta \ln\left(\frac{\Delta}{a}\right) \qquad (4.101)$$

A similar procedure is applied for the remaining modes to obtain

$$\frac{E_{zn}}{H_{\varphi n}}\Big|_{z=\Delta} \simeq \frac{j\omega\mu_0}{n} \Delta \frac{\Delta^{2n} - a^{2n}}{\Delta^{2n} + a^n}, \qquad n \neq 0 \qquad (4.102)$$

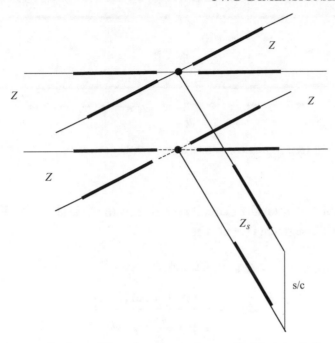

FIGURE 4.20: Structure of a shunt 2D node representing a cell containing a fine wire

We will now devise a circuit which presents to mode $n = 0$ the impedance given by (4.101) and to the other modes the impedance given by (4.102). The topology of the circuit is shown in Fig. 4.20 where the link lines have a characteristic impedance Z and the short-circuit stub an impedance Z_s. Using the modal structure in Table 4.1 it can be shown that for modes $n \neq 0$ the voltage at the center of the node is equal to zero. Hence, these modes see effectively a short-circuited TL segment of characteristic impedance Z and length Δ. We demand that the input impedance of this segment is equal to the value given by (4.102).

$$Z_{\text{IN}} = j Z \tan(k_{\text{TL}} \Delta) = \frac{j \omega \mu_0}{n} \Delta \frac{\Delta^{2n} - a^{2n}}{\Delta^{2n} + a^{2n}}, \qquad n \neq 0 \qquad (4.103)$$

where the propagation constant is given $k_{\text{TL}} = \omega \sqrt{\mu_0 \varepsilon_0}/\sqrt{2}$. Substituting and approximating $\tan(.)$ we obtain

$$Z = Z_{\text{TL}} \frac{1}{n} \frac{\Delta^{2n} - a^{2n}}{\Delta^{2n} + a^{2n}}, \qquad n \neq 0 \qquad (4.104)$$

where $Z_{\text{TL}} = \sqrt{2} \left(\sqrt{\mu_0/\varepsilon_0} \right)$ is the link line impedance of the nodes representing free space in the absence of an embedded wire. This choice of Z ensures that mode $n = 1$ sees the correct impedance. Mode $n = 2$, as before, is not presented with exactly the correct impedance. We must now ensure that mode $n = 0$ sees the correct impedance. This is the only mode that is

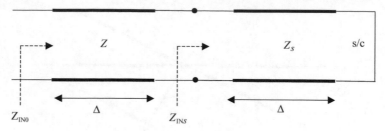

FIGURE 4.21: Circuit seen by mode $n = 0$ (see Fig. 4.20)

affected by the stub! The equivalent circuit seen by this mode is shown in Fig. 4.21. From the cascade of the two TL segments we obtain

$$Z_{INs} = j Z_s \tan (k_{TL}\Delta)$$

$$Z_{IN0} = \frac{\frac{Z_{INs}}{Z} + j \tan (k_{TL}\Delta)}{1 + j\frac{Z_{INs}}{Z} \tan (k_{TL}\Delta)}$$

Combining these two equations and demanding that Z_{IN0} is equal to the expression in (4.101) we obtain the required stub characteristic impedance,

$$Z_s = Z_{TL} \ln \left(\frac{\Delta}{a}\right) - Z \qquad (4.105)$$

Therefore, the presence of a wire inside a node is recognized by altering the parameters of the node as shown in Fig. (4.20) and Eqs. (4.104) and (4.105). Pulses coming from the free space region, upon encountering the boundary of the node containing the wire, experience a discontinuity (coming from an impedance Z_{TL} to an impedance Z). Scattering in the wire node is also different because of the presence of the stub. The radius of the wire can be much smaller that the radius of the node ($a \ll \Delta$). In this way the size of the computation is not dictated by the size of the wire. Yet its impact is much more accurately included in the model. Classical approaches to this problem are based on the formulation given in [20]. Here the local solution embedded into the model is based on the quasi-static fields around the wire, i.e., the magnetic and electric fields near a wire carrying current I and charge Q are given by

$$H(r) = {I}/{2\pi r}, \qquad E(r) = {Q}/{2\pi \varepsilon r} \qquad (4.106)$$

One should contrast Eq. (4.106) with (4.98) and (4.99). In the modal expansion technique (MET) adopted here we take account of four modes (in 2D), while in the classical approach that of only one. We therefore get a greater accuracy and we are also able to locate the wire anywhere inside the node (not just centrally). This is achieved by exploiting the addition theorems for Bessel functions. Further details on the MET and how it is applied for offset wires and multiconductor systems (more than one wire inside a node) may be found in [21–23]. Features other than wires may also be embedded inside a node such as dielectric rods, etc., using a similar approach [24].

CHAPTER 5

An Unstructured 2D TLM Model

The development of TLM models we have described so far is based on highly structured meshes which are aligned along Cartesian coordinates. It is possible to develop meshes based on cylindrical or spherical coordinates to suit problems with a particular symmetry and these are available in TLM [25, 26]. However, whatever the coordinate system chosen, these remain highly structured meshes. This makes it easy to establish a mesh and because of the high level of regularity (most nodes have exactly the same parameters) programing and tracking various quantities is simple. It is nevertheless true that a structured mesh is not ideally suited to describing boundaries which do not coincide with coordinate axes, e.g., curved boundaries in a Cartesian mesh. This is illustrated in Fig. 5.1(a) where we see that a curved boundary is approximated by a "stair-cased" boundary resulting under certain circumstances in stair-casing errors [27]. In contrast, if instead of a rectangle as the basic nodal element we employ a triangle as shown in Fig. 5.1(b) then a much improved description of the boundary is achieved. We now have, however, a collection of triangles of different shapes and sizes forming an unstructured mesh. The disadvantage is that each node is potentially different thus requiring storage of a substantial amount of node-based information. In addition, control of the time step to maintain synchronism across the mesh is now not a trivial matter. The accuracy of the results is strongly dependent on the "quality" of the unstructured mesh. All these difficulties are familiar to researchers of the Finite Element Method where traditionally unstructured meshes are employed. As is common in modeling, an unstructured mesh solves some problems and creates others and it is up to the modeler to reach a balanced judgement as to where and how an unstructured mesh should be employed.

We describe below the theoretical development of 2D TLM unstructured models in the time-domain.

5.1 A TRIANGULAR MESH IN TLM

We focus on a shunt node for TM waves shown schematically in Fig. 5.2 [28]. The figure shows three ports connecting the node to its neighbors all originating from the center of the node. The full geometrical description is contained in the distances Δ_i and angles ϕ_i shown. The electric field E_z is perpendicular to the plane of the paper.

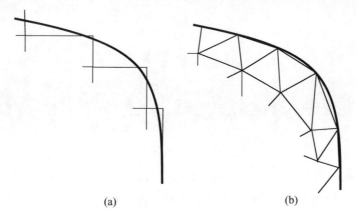

(a)   (b)

FIGURE 5.1: Curved boundary approximation using a Cartesian mesh (a) and a triangular mesh (b)

The general approach, as before, is to decompose the electric field into modal components (three in this case as we have three degrees of freedom) and work out an admittance matrix relating magnetic and electric fields. After that this admittance matrix is mapped onto a circuit to obtain the TLM model.

We express the fields in the vicinity of the node as the superposition of the first three modes of an expansion of the field in cylindrical coordinates around the node center. This gives

$$E_z(r, \vartheta) = J_0(kr)X_{c0} + \cos(\vartheta)J_1(kr)\frac{2X_{c1}}{k} + \sin(\vartheta)J_1(kr)\frac{2X_{s1}}{k} \tag{5.1}$$

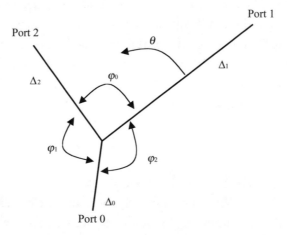

FIGURE 5.2: The basic notation and structure of a triangular node

where the coefficients X represent the modal mix of the total field. Similarly, for the magnetic field we have

$$-j\omega\mu_0 H_\vartheta(r, \vartheta) = \frac{\partial E_z(r, \vartheta)}{\partial r} \tag{5.2}$$

The electric field at the three ports may now be expressed in terms of the modal components X by applying (5.1) at each port. As an example, for port 1 we get

$$E_{z1} = J_0(k\Delta_1)X_{c0} + \cos(0)J_1(k\Delta_1)\frac{2X_{c1}}{k} + \sin(0)J_1(k\Delta_1)\frac{2X_{s1}}{k}$$

$$\simeq \left[1 - \frac{(k\Delta_1)^2}{4}\right]X_{c0} + \frac{k\Delta_1}{2}\frac{2X_{c1}}{k} + 0X_{s1} \tag{5.3}$$

$$\simeq X_{c0} + \Delta_1 X_{c1}$$

where we have used small argument approximations for the Bessel functions and neglected terms of the order $(k\Delta)^2$. Equation (5.3) gives the first row of a matrix $\boldsymbol{T_e}$ relating the electric field to its modal components.

$$\boldsymbol{E_z} = \boldsymbol{T_e}\boldsymbol{X} \tag{5.4}$$

where

$$\boldsymbol{E_z} = [\,E_{z1}\quad E_{z2}\quad E_{z0}\,]^T, \qquad \boldsymbol{X} = [\,X_{c0}\quad X_{c1}\quad X_{s1}\,]^T$$

$$\boldsymbol{T_e} = \begin{bmatrix} 1 & \Delta_1 & 0 \\ 1 & \Delta_2\cos(\varphi_0) & \Delta_2\sin(\varphi_0) \\ 1 & \Delta_0\cos(\varphi_0 + \varphi_1) & \Delta_0\sin(\varphi_0 + \varphi_1) \end{bmatrix} \tag{5.5}$$

We express in a similar way the magnetic field using (5.2)

$$-j\omega\mu_0 H_\vartheta(r, \vartheta) = \frac{\partial J_0(kr)}{\partial r}X_{c0} + \frac{\partial J_1(kr)}{\partial r}\cos(\vartheta)\frac{2X_{c1}}{k} + \frac{\partial J_1(kr)}{\partial r}\sin(\vartheta)\frac{2X_{s1}}{k} \tag{5.6}$$

The derivatives of the Bessel functions are then replaced by their small argument approximations (Appendix 2) to obtain the magnetic field in port 1.

$$-j\omega\mu_0 H_{\vartheta1} = -\frac{k^2\Delta_1}{2}X_{c0} + \frac{2X_{c1}}{k}\cos(0)\frac{k}{2} + \frac{2X_{s1}}{k}\sin(0)\frac{k}{2}$$

$$-j\omega\mu_0\Delta_1 H_{\vartheta1} = -\frac{k^2\Delta_1^2}{2}X_{c0} + \Delta_1 X_{c1} \tag{5.7}$$

The last expression is effectively the first row of the matrix T_h relating the magnetic field vector to its modal components.

$$j\omega\mu_0\Delta^D H_\vartheta = T_h X \tag{5.8}$$

where the transformation matrix is

$$T_h = \begin{bmatrix} -\dfrac{(k\Delta_1)^2}{2} & \Delta_1 & 0 \\[2mm] -\dfrac{(k\Delta_2)^2}{2} & \Delta_2\cos(\varphi_0) & \Delta_2\sin(\varphi_0) \\[2mm] -\dfrac{(k\Delta_0)^2}{2} & \Delta_0\cos(\varphi_0+\varphi_1) & \Delta_0\sin(\varphi_0+\varphi_1) \end{bmatrix} \tag{5.9}$$

$$\Delta^D = \begin{bmatrix} \Delta_1 & & \\ & \Delta_2 & \\ & & \Delta_0 \end{bmatrix} \tag{5.10}$$

From (5.4) we obtain $X = T_e^{-1}E_z$ and substituting in (5.8) for X we get

$$j\omega\mu_0\Delta^D H_\vartheta = T_h T_e^{-1}E_z \tag{5.11}$$

Inverting the matrix T_e and carrying out the manipulations in (5.11) we obtain an expression relating the magnetic field in the three ports to the electric field, i.e., the elements of an admittance operator. This is written out in full below.

$$\begin{bmatrix} H_{\vartheta 1} \\ H_{\vartheta 1} \\ H_{\vartheta 1} \end{bmatrix} = j\omega\frac{\varepsilon_0}{2} \frac{\begin{bmatrix} \Delta_2\Delta_0\Delta_1 s_1 & \Delta_1\Delta_0\Delta_1 s_2 & \Delta_2\Delta_1\Delta_1 s_0 \\ \Delta_2\Delta_0\Delta_2 s_1 & \Delta_2\Delta_0\Delta_1 s_2 & \Delta_2\Delta_2\Delta_1 s_0 \\ \Delta_2\Delta_0\Delta_0 s_1 & \Delta_0\Delta_0\Delta_1 s_2 & \Delta_2\Delta_0\Delta_1 s_0 \end{bmatrix}}{\Delta_2\Delta_0 s_1 + \Delta_1\Delta_0 s_2 + \Delta_2\Delta_1 s_0} \begin{bmatrix} E_{z1} \\ E_{z2} \\ E_{z0} \end{bmatrix}$$

$$+\frac{1}{j\omega\mu_0} \frac{\begin{bmatrix} \Delta_0 s_2 + \Delta_2 s_0 & -\Delta_0 s_2 & -\Delta_2 s_0 \\ -\Delta_0 s_1 & \Delta_0 s_1 + \Delta_1 s_0 & -\Delta_1 s_0 \\ -\Delta_2 s_1 & -\Delta_1 s_2 & \Delta_1 s_2 + \Delta_2 s_1 \end{bmatrix}}{\Delta_2\Delta_0 s_1 + \Delta_1\Delta_0 s_2 + \Delta_2\Delta_1 s_0} \begin{bmatrix} E_{z1} \\ E_{z2} \\ E_{z0} \end{bmatrix} \tag{5.12}$$

where $s_i = \sin(\varphi_i)$. Examining (5.12) we can see that the first term on the RHS is capacitive and the second inductive. This gives us some encouragement that an L–C network may be constructed to present an admittance in accordance with (5.12). To investigate further we map electric fields to voltages and magnetic fields to currents as shown below.

$$
\begin{bmatrix} E_{z1} \\ E_{z2} \\ E_{z0} \end{bmatrix} \rightarrow \begin{bmatrix} V_1 \\ V_2 \\ V_0 \end{bmatrix}, \qquad \begin{bmatrix} \alpha s_1 & 0 & 0 \\ 0 & \alpha s_2 & 0 \\ 0 & 0 & \alpha s_0 \end{bmatrix} \begin{bmatrix} H_{\vartheta 1} \\ H_{\vartheta 2} \\ H_{\vartheta 0} \end{bmatrix} \rightarrow \begin{bmatrix} I_1 \\ I_2 \\ I_0 \end{bmatrix} \tag{5.13}
$$

where α is a constant to be determined. Substituting (5.13) into (5.12) we obtain an equation relating port voltages and currents.

$$
\begin{bmatrix} I_1 \\ I_2 \\ I_0 \end{bmatrix} = j\omega\alpha \frac{\varepsilon_0}{2} \frac{\begin{bmatrix} \Delta_2\Delta_0\Delta_1 s_1 s_1 & \Delta_1\Delta_0\Delta_1 s_2 s_1 & \Delta_2\Delta_1\Delta_1 s_0 s_1 \\ \Delta_2\Delta_0\Delta_2 s_1 s_2 & \Delta_2\Delta_0\Delta_1 s_2 s_2 & \Delta_2\Delta_2\Delta_1 s_0 s_2 \\ \Delta_2\Delta_0\Delta_0 s_1 s_0 & \Delta_0\Delta_0\Delta_1 s_2 s_0 & \Delta_2\Delta_0\Delta_1 s_0 s_0 \end{bmatrix}}{\Delta_2\Delta_0 s_1 + \Delta_1\Delta_0 s_2 + \Delta_2\Delta_1 s_0} \begin{bmatrix} V_1 \\ V_2 \\ V_0 \end{bmatrix}
$$

$$
+ \frac{\alpha}{j\omega\mu_0} \frac{\begin{bmatrix} \Delta_0 s_2 s_1 + \Delta_2 s_0 s_1 & -\Delta_0 s_2 s_1 & -\Delta_2 s_0 s_1 \\ -\Delta_0 s_1 s_2 & \Delta_0 s_1 s_2 + \Delta_1 s_0 s_2 & -\Delta_1 s_0 s_2 \\ -\Delta_2 s_1 s_0 & -\Delta_1 s_2 s_0 & \Delta_1 s_2 s_0 + \Delta_2 s_1 s_0 \end{bmatrix}}{\Delta_2\Delta_0 s_1 + \Delta_1\Delta_0 s_2 + \Delta_2\Delta_1 s_0} \begin{bmatrix} V_1 \\ V_2 \\ V_0 \end{bmatrix} \tag{5.14}
$$

A circuit topology with three ports such that the port currents and voltages are related as shown in (5.14) would give us the desired model. However, there is a problem. Circuits consisting of inductors, capacitors, etc., are reciprocal [29]—the admittance matrix is symmetric. Examining (5.14) we see that the inductive term is reciprocal but this is not the case for the capacitive term. In order to get a physically realizable circuit we need to have symmetrical matrices. This requirement invites further examination of the matrices to see whether reciprocity can be achieved. It is evident from (5.14) that the inductive term dominates at low frequencies (the domain of model validity). However, we cannot altogether neglect the capacitive term as in the case when all port voltages are the same the inductive term vanishes. This can be confirmed by direct substitution into the second term on the RHS with $V_1 = V_2 = V_0$. Under these circumstances, the only term left is the capacitive one and it is equal to

$$
\begin{bmatrix} I_1 \\ I_2 \\ I_0 \end{bmatrix} = j\omega\alpha \frac{\varepsilon_0}{2} \begin{bmatrix} \Delta_1 s_1 & 0 & 0 \\ 0 & \Delta_2 s_2 & 0 \\ 0 & 0 & \Delta_0 s_0 \end{bmatrix} \begin{bmatrix} V_1 \\ V_2 \\ V_0 \end{bmatrix}, \qquad \omega << \tag{5.15}
$$

We therefore replace the capacitive term in (5.14) with the form given by (5.15) that is reciprocal.

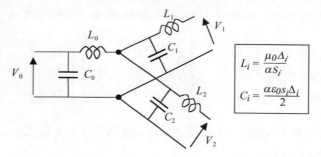

FIGURE 5.3: Network representation for a triangular-shaped cell

The expression we now need to map onto a circuit is given below.

$$
\begin{bmatrix} I_1 \\ I_2 \\ I_0 \end{bmatrix} \simeq j\omega\alpha\frac{\varepsilon_0}{2} \begin{bmatrix} \Delta_1 s_1 & 0 & 0 \\ 0 & \Delta_2 s_2 & 0 \\ 0 & 0 & \Delta_0 s_0 \end{bmatrix} \begin{bmatrix} V_1 \\ V_2 \\ V_0 \end{bmatrix}
$$

$$
+\frac{\alpha}{j\omega\mu_0} \frac{\begin{bmatrix} \Delta_0 s_2 s_1 + \Delta_2 s_0 s_1 & -\Delta_0 s_2 s_1 & -\Delta_2 s_0 s_1 \\ -\Delta_0 s_1 s_2 & \Delta_0 s_1 s_2 + \Delta_1 s_0 s_2 & -\Delta_1 s_0 s_2 \\ -\Delta_2 s_1 s_0 & -\Delta_1 s_2 s_0 & \Delta_1 s_2 s_0 + \Delta_2 s_1 s_0 \end{bmatrix}}{\Delta_2 \Delta_0 s_1 + \Delta_1 \Delta_0 s_2 + \Delta_2 \Delta_1 s_0} \begin{bmatrix} V_1 \\ V_2 \\ V_0 \end{bmatrix}
$$

(5.16)

The circuit representing this admittance matrix is shown in Fig. 5.3. This can be confirmed directly by applying nodal analysis in the circuit to relate voltages and currents. We note that the circuit parameters are given by

$$
L_i = \frac{\mu_0 \Delta_i}{a s_i}, \qquad C_i = \frac{\alpha \varepsilon_0 s_i \Delta_i}{2}
$$

(5.17)

In order for these components to be positive (to ensure stability) $s_i = \sin(\varphi_i)$ must be positive and hence all the angles ϕ_i must be less than π. This constraint is familiar to finite element practitioners where similar meshes are used and is known as Delaunay triangulation [30]. It maximizes the minimum angles of all the triangles in an effort to avoid sliver (long and thin) triangles. In FE work a sliver triangle whose two sides are almost identical indicates solutions which are almost the same and hence a system of equations containing two almost identical equations. This gives ill-conditioned matrices making a numerical solution difficult and inaccurate. In unstructured TLM, the Delaunay condition ensures positive passive element and hence stability.

A further condition to be investigated is the continuity of electric and magnetic fields across node boundaries. We see from (5.13) that continuity of voltage at the boundary between nodes ensures continuity of electric field. However, continuity of current does not automatically ensure continuity of H since $I_i = \alpha s_i H_i$. In order to get magnetic field continuity we need to enforce the condition that αs_i is the same in adjacent nodes. A full discussion of this may be found in [28] and it leads to two possibilities. The first is to define the node center as the center of gravity (COG) of the triangle defined by its ports. The centroid (COG) of a triangle is the point at which the medians of the triangle intersect. The second approach is to define node centers as the circumcenters (CCM) of the Delaunay triangulation. We discuss further only the CCM with the definition of the node center shown in Fig. 5.4. From this figure we obtain, $2Rs_i = \ell_i$ where R is the circumradius of the triangle and ℓ_i is the side of the triangle opposite to the angle ϕ_i. If we choose the constant α to be equal to $2R$ for each node, then this gives for each node, $I_i = as_i H_i = 2Rs_i H_i = \ell_i H_i$. Since the length ℓ_i is the same across the sides of adjacent nodes it follows that continuity of current also means continuity of magnetic field. The ports for the CCM configuration lie halfway between the node centers as shown in Fig. 5.4. Unstructured meshes need careful examination to obtain optimum conditions especially as regards the permissible time step. For computational efficiency reasons it is important to maximize the time step. This, in a time-domain method, means that the link line length must not be allowed to become too small. Even in a Delaunay mesh there are circumstances where the circumcenter is outside the triangle and may be very close (or even coincide) to the circumcenter of another triangle. Thus, this means that the link line between these two circumcenters becomes very short with very unfavorable implications for the time step. Such conditions must be detected and avoided. One approach is to merge such triangles into a quadrilateral. Details of these techniques may be found in [28].

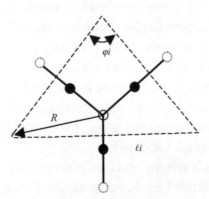

FIGURE 5.4: Node center defined as the circumcenter (inner small circle), centers of adjacent nodes (outer dotted small circles) and node interfaces (full small circles)

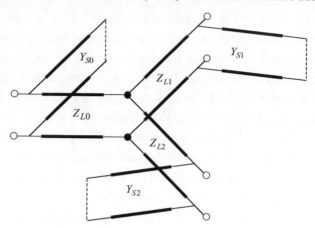

FIGURE 5.5: TLM representation of a triangular cell

The implementation of the circuit shown in Fig. 5.3 in TLM is straightforward. One option is to model all inductance in the link lines and any deficit in capacitance is added with stubs as shown in Fig. 5.5. Scattering and connection are implemented following the same principles as for the rectangular shunt node.

5.2   APPLICATIONS OF THE TRIANGULAR MESH

I have indicated at the beginning of this chapter the advantages and disadvantages of unstructured meshes. The development in Section 5.1 will have made it clearer how and when to use a triangular mesh. There is a significant overhead as each node is different and the quality of the mesh has a significant impact on accuracy. Controlling the time step to avoid pathological conditions such as those arising when two circumcenters coincide or are very near to each other is imperative. Extensive stubbing may be required to maintain synchronism and this again adds to dispersion. Each Delaunay mesh will give a solution with a different error. It appears sensible to suggest that unstructured meshes should be used only when necessary. I can see two main areas where they may be employed with a clear advantage.

First, in case of intricate or curved boundaries, where the avoidance of stair-casing errors is important, using triangular meshes may be the best option. But even then consideration must be given as to whether the entire space must be meshed with triangles instead of only areas near the boundary to conform the mesh better to it.

Second, a useful approach in many problems is to use a hybrid mesh where certain regions are meshed with a coarse structured mesh, others with a fine structured mesh and some with an unstructured triangular mesh. In this way an optimum mesh is employed for each particular part of the problem. The triangular mesh can be used to "stitch" together different mesh areas as

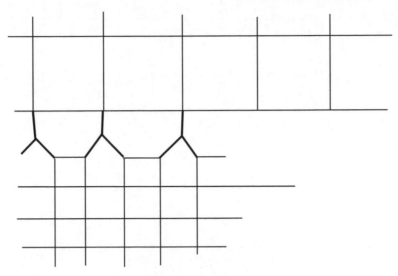

FIGURE 5.6: Triangular elements (shown in thicker line) used to join two Cartesian meshes of different spatial resolutions

shown schematically in Fig. 5.6. This technique is described further in [31]. Therefore, rather than an indiscriminate use of an unstructured mesh a more intelligent approach is to use it as a high-quality, specialist, expensive modeling medium. Used in this way it is an excellent complement to other meshes in TLM.

CHAPTER 6

TLM in Three Dimensions

6.1 BASIC CONCEPTS IN 3D TLM

We have traced the development of TLM models for structured and unstructured meshes in 2D and we must now examine how these techniques are applied in three dimensions (3D). Naturally, 3D networks need to be devised to do the necessary mapping. The approach is very similar to that adopted in 2D but with extra complexity due to the higher dimensionality. I will therefore avoid excessive detail that does not add much to the physical understanding and the modeling philosophy that was already described as part of the 2D work. I will focus instead in this section on how modeling is done in 3D and leave more sophisticated topics for the following sections and for self-study of the available literature.

We start again with a cell of space with dimensions $\Delta x \times \Delta y \times \Delta z$ and seek to represent the properties of the EM in this cell by analogy to the behavior of a node in a 3D network. Over the years, a number of nodes have been developed for this purpose but the most widely used structure at present is the Symmetrical Condensed Node (SCN) shown in Fig. 6.1 [32]. It consists of a network of interconnected lines such that on each face of the cell there correspond two ports orthogonal to each other. In this way, any field polarization can be accounted for. The port voltages are labeled by numbers 1–12 in the traditional labeling scheme or by three subscripts, the first indicating the direction of propagation (x, y, or z), the second indicating whether the segment is along the negative or positive coordinate axis (n or p), and the third the polarization of the pulse (x, y, or z). A superscript may be added to indicate incident or reflected pulse (i or r). In addition stubs may be added to account for different materials, noncubical cell shapes, and losses. Rather than complicating matters from the start, we will examine first the simple cubic SCN representing an air-filled cell to get the basic scattering and connection schemes. We will also postpone a discussion of the parameters of the node to map a particular cell to another section as this is intricately connected with material properties, shape of the cell, and the use of stubs. As in the case of the 2D nodes we need to address scattering, connection, excitation and output, and treatment of boundaries. However, unlike in the case of the 2D nodes a simple equivalent circuit for the SCN cannot be derived and it is therefore necessary to obtain its scattering properties by resorting to more general principles. Examining carefully the structure of the SCN in Fig. 6.1 you may identify "series" and "shunt" nodes as its constituents.

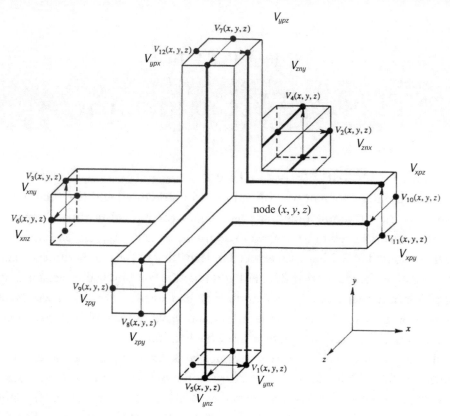

FIGURE 6.1: The 3D Symmetrical Condensed Node (SCN) with two alternative port notations

Two such nodes are shown in Fig. 6.2 where for generality I have also indicated the characteristic impedance of each line segment to be different. It would be a mistake to suppose that V_y may be calculated from Fig. 6.2(a) as for the 2D shunt node and therefore that $V_{xny}^r = V_y - V_{xny}^i$. This expression is incorrect for the 3D node. I have taken care to indicate in Fig. 6.2 that a connection at the center of these two nodes is somehow implied but does not actually exist. An equivalent voltage V_y and current I_z may be defined but they are not the actual voltage and current on each line. We will return to this idea later on. If a simple circuit cannot be derived how the scattering matrix can be obtained? The answer lies in enforcing fundamental principles [32, 33, 34]. Specifically, we demand

- Conservation of electric charge.
- Conservation of magnetic flux.
- Electric field continuity.
- Magnetic field continuity.

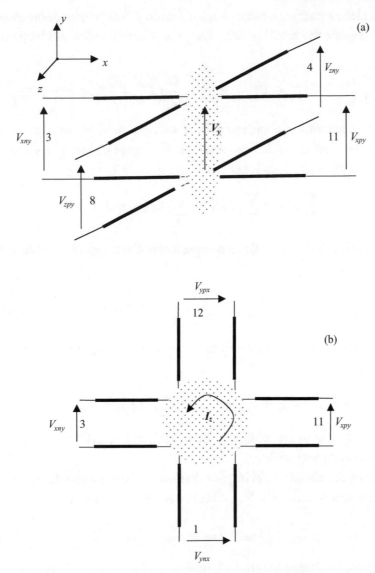

FIGURE 6.2: Shunt (a) and series (b) "nodes" extracted from Fig. 6.1 used to illustrate scattering in the 3D SCN

Enforcement of each of these principles gives three equations relating incident and reflected voltages and therefore 12 equations in total. This is enough to calculate the elements of the scattering matrix relating the 12 reflected voltages components to the 12 incident components. We assume for simplicity that the total capacitance and inductance of each link line are C and L, respectively. We enforce each condition separately.

Following electric charge conservation law for the y-component of the electric field shown in Fig. 6.2(a) we equate the total incident charge to the total reflected charge for the four half link –lines

$$\frac{C}{2}\left(V^i_{xny} + V^i_{zny} + V^i_{xpy} + V^i_{zpy}\right) = \frac{C}{2}\left(V^r_{xny} + V^r_{zny} + V^r_{xpy} + V^r_{zpy}\right) \tag{6.1}$$

Other two equations are obtained in a similar number for the x-component and z-component of the electric field. Conservation of magnetic flux $\left(\oiint \boldsymbol{B}\cdot d\boldsymbol{s} = 0\right)$ implies that the total flux linked to all the lines is zero, i.e.,

$$\sum_n \lambda_n = \sum_n L_n I_n = \sum_n L_n \left(I^i_n - I^r_n\right) = 0 \tag{6.2}$$

Expressing this condition for the z-component of the magnetic field shown in Fig. 6.2(b) we get

$$\frac{L}{2}\left(-I^i_{xny} - I^i_{ypx} + I^i_{xpy} + I^i_{ynx}\right) = \frac{L}{2}\left(-I^r_{xny} - I^r_{ypx} + I^r_{xpy} + I^r_{ynx}\right) \tag{6.3}$$

Substituting in (6.3) in terms of the voltage and recognizing that $I^i = V^i/Z$, $I^r = -V^r/Z$ we obtain

$$-V^i_{xny} - V^i_{ypx} + V^i_{xpy} + V^i_{ynx} = V^r_{xny} + V^r_{ypx} - V^r_{xpy} - V^r_{ynx} \tag{6.4}$$

Other two equations are obtained in a similar manner for the x-component and y-component of the magnetic field.

Continuity of the electric field implies that the y-component whether calculated on the z-directed or the x-directed lines in Fig. 6.2(a) must be the same, i.e.,

$$V_{zpy} + V_{zny} = V_{xpy} + V_{xny}$$

or, expressing the total voltages in terms of incident and reflected voltages we get

$$V^i_{zpy} + V^r_{zpy} + V^i_{zny} + V^r_{zny} = V^i_{xpy} + V^r_{xpy} + V^i_{xny} + V^r_{xny} \tag{6.5}$$

Other two equations are obtained in a similar number for the x-component and z-component of the electric field.

Continuity of the magnetic field implies that the z-component must be the same whether calculated from the x-directed or y-directed lines in Fig. 6.2(b), i.e.,

$$I_{xpy} - I_{xny} = I_{ynx} - I_{ypx}$$

or, expressing the currents in terms of incident and reflected voltages we get

$$V_{xpy}^i - V_{xpy}^r - (V_{xny}^i - V_{xny}^r) = V_{ynx}^i - V_{ynx}^r - (V_{ypx}^i - V_{ypx}^r) \tag{6.6}$$

Other two equations are obtained in a similar manner for the x-component and y-component of the magnetic field.

Equations (6.1), (6.4–6.6) and their equivalents for the remaining polarizations form a system of 12 independent equations that give all the information necessary to derive scattering for the SCN

$$_k V^r = S_k V^i \tag{6.7}$$

where the scattering matrix for the SCN representing a cubic cell is

$$
S = 0.5
\begin{bmatrix}
0 & 1 & 1 & 0 & 0 & 0 & 0 & 0 & 1 & 0 & -1 & 0 \\
1 & 0 & 0 & 0 & 0 & 1 & 0 & 0 & 0 & -1 & 0 & 1 \\
1 & 0 & 0 & 1 & 0 & 0 & 0 & 1 & 0 & 0 & 0 & -1 \\
0 & 0 & 1 & 0 & 1 & 0 & -1 & 0 & 0 & 0 & 1 & 0 \\
0 & 0 & 0 & 1 & 0 & 1 & 0 & -1 & 0 & 1 & 0 & 0 \\
0 & 1 & 0 & 0 & 1 & 0 & 1 & 0 & -1 & 0 & 0 & 0 \\
0 & 0 & 0 & -1 & 0 & 1 & 0 & 1 & 0 & 1 & 0 & 0 \\
0 & 0 & 1 & 0 & -1 & 0 & 1 & 0 & 0 & 0 & 1 & 0 \\
1 & 0 & 0 & 0 & 0 & -1 & 0 & 0 & 0 & 1 & 0 & 1 \\
0 & -1 & 0 & 0 & 1 & 0 & 1 & 0 & 1 & 0 & 0 & 0 \\
-1 & 0 & 0 & 1 & 0 & 0 & 0 & 1 & 0 & 0 & 0 & 1 \\
0 & 1 & -1 & 0 & 0 & 0 & 0 & 0 & 1 & 0 & 1 & 0
\end{bmatrix}
\tag{6.8}
$$

and the 12 vectors for incident and reflected pulses follow the number labeling of Fig. 6.1. When programing scattering for computation we rarely use the matrix given in (6.8). Instead, it is more efficient to calculate the reflected voltages directly from the following equations (using the alternative labeling).

$$V_{ynx}^r = \frac{1}{2}\left(V_{znx}^i + V_{zpy}^i + V_{xny}^i - V_{xpy}^i\right)$$

$$V_{ypx}^r = \frac{1}{2}\left(V_{znx}^i + V_{zpx}^i + V_{xpy}^i - V_{xny}^i\right)$$

$$V_{znx}^r = \frac{1}{2}\left(V_{ynx}^i + V_{ypx}^i + V_{xnz}^i - V_{xpz}^i\right)$$

$$V_{zpx}^r = \frac{1}{2}\left(V_{ynx}^i + V_{ypx}^i + V_{xpz}^i - V_{xnz}^i\right)$$

$$V_{zny}^r = \frac{1}{2}\left(V_{xny}^i + V_{xpy}^i + V_{ynz}^i - V_{ypz}^i\right)$$

$$V_{zpy}^r = \frac{1}{2}\left(V_{xny}^i + V_{xpy}^i + V_{ypz}^i - V_{ynz}^i\right)$$

$$V_{xny}^r = \frac{1}{2}\left(V_{zny}^i + V_{zpy}^i + V_{ynx}^i - V_{ypx}^i\right)$$

$$V_{xpy}^r = \frac{1}{2}\left(V_{zny}^i + V_{zpy}^i + V_{ypx}^i - V_{ynx}^i\right)$$

$$V_{xnz}^r = \frac{1}{2}\left(V_{ynz}^i + V_{ypz}^i + V_{znx}^i - V_{zpx}^i\right) \qquad (6.9)$$

$$V_{xpz}^r = \frac{1}{2}\left(V_{ynz}^i + V_{ypz}^i + V_{zpx}^i - V_{znx}^i\right)$$

$$V_{ynz}^r = \frac{1}{2}\left(V_{xnz}^i + V_{xpz}^i + V_{zny}^i - V_{zpy}^i\right)$$

$$V_{ypz}^r = \frac{1}{2}\left(V_{xnz}^i + V_{xpz}^i + V_{zpy}^i - V_{zny}^i\right)$$

The scattering in Eqs. (6.8) or (6.9) is based on a cubical cell ($\Delta x = \Delta y = \Delta z = \Delta \ell$), where all the link lines have the same characteristic impedance and from (3.8) and (3.9) $\Delta t = \sqrt{LC}$, $Z_0 = \sqrt{L/C}$. From (3.10) we also get $C = \Delta t / Z_0$, $L = \Delta t Z_0$. Capacitance and inductance must correspond to those of the cell representing the block of medium with parameters ε, μ. For example, the x-directed capacitance of the cell must be represented by the four half lines with fields polarized in the x-direction, i.e.,

$$\varepsilon \frac{\Delta y \Delta z}{\Delta x} = \varepsilon \Delta \ell = 4\frac{\Delta t /2}{Z_0} = \frac{2\Delta t}{Z_0} \qquad (6.10)$$

Similarly, the inductance represented by the four half lines must be

$$\mu \frac{\Delta y \Delta z}{\Delta x} = \mu \Delta \ell = 4\frac{\Delta t}{2}Z_0 = 2\Delta t Z_0 \qquad (6.11)$$

Multiplying (6.10) by (6.11) to eliminate Z_0 we then obtained the required time step to represent this block of medium

$$\Delta t = \frac{\Delta \ell}{2u} \qquad (6.12)$$

where $u = 1/\sqrt{\mu \varepsilon}$. For a cell in free space $u = c$ the speed of light.

We have described so far the basic scattering procedure in 3D TLM. Connection proceeds in exactly the same way as for the 2D nodes by exchanging pulses with neighbors. Computation procedures in 3D TLM are identical to those in 2D except for the fact that we now deal with 12 rather than 4 pulses.

This basic introduction would be incomplete without a mention of how we deal with inhomogeneous materials and noncubical cells. We had a similar discussion in the previous chapter regarding inhomogeneous media. Since the velocity of propagation changes, it appears that we have two options. First, we may keep $\Delta\ell$ constant and thus we have one-to-one correspondence across material boundaries but then Δt will be different in different materials and we will loose synchronism. Second, we may maintain synchronism but loose one-to-one correspondence. Neither of these alternatives is attractive. Instead, we follow a procedure where link lines represent a background medium (free space in most cases) and we add capacitive stubs to account for $\varepsilon_r > 1$ and inductive stubs for $\mu_r > 1$. In this way, we maintain both synchronism and one-to-one correspondence across material boundaries. The other desirable feature of being able to alter the shape of a cell from a cube to a general cuboid shape is dealt exactly in the same way by introducing stubs to model C and L correctly in the different directions. The only aspect to watch out for is that we keep stub parameters positive to ensure stability. The principles are straightforward. If we wish to represent a material with $\varepsilon_r > 1$ in a TLM model where the background material is free space, then in the x-direction the desired capacitance is made out of the capacitance of the link lines [see (6.10)] plus the capacitance of an open-circuit stub.

$$\varepsilon\Delta\ell = \frac{2\Delta t}{Z_0} + C_{ox} \tag{6.13}$$

Hence, substituting $\varepsilon_0 = 1/(Z_0 c)$ in (6.13) we obtain

$$C_{ox} = \frac{1}{Z_0}(\frac{\varepsilon_r}{c}\Delta\ell - 2\Delta t) \tag{6.14}$$

Taking the round-trip time for the stub to be Δt we then obtain its characteristic admittance

$$Y_{ox} = \frac{2C_{ox}}{\Delta t} = \frac{2}{Z_0}\left(\frac{\varepsilon_r}{c}\frac{\Delta\ell}{\Delta t} - 2\right) = \frac{1}{Z_0}4(\varepsilon_r - 1) \tag{6.15}$$

where for stability the time step based on the material with the lowest dielectric permittivity (free space) has been chosen.

The presence of materials with different permeability is dealt with in a similar manner by introducing inductive (short-circuit) stubs. The characteristic impedance of an inductive stub is then,

$$Z_{sx} = Z_0 4(\mu_r - 1) \tag{6.16}$$

Different stubs may be introduced in other directions to account for the anisotropic properties.

Losses may be introduced by inserting an appropriate conductance. For a medium with a complex dielectric constant given by

$$\varepsilon^* = \varepsilon_r \varepsilon_0 - j\varepsilon'' = \varepsilon_r \varepsilon_0 \left(1 - j\frac{\varepsilon''}{\varepsilon_r \varepsilon_0} \right) \qquad (6.17)$$

and conduction conductivity σ, the total effective electric conductivity may be defined as

$$\sigma_e = \omega\varepsilon'' + \sigma \qquad (6.18)$$

Equation (6.18) may be written in terms of the loss tangent for the material that incorporates in an approximate manner the conduction conductivity

$$\varepsilon^* = \varepsilon_r \varepsilon_0 (1 - j\tan\delta) \qquad (6.19)$$

where $\tan\delta = \sigma_e/(\omega\varepsilon_r\varepsilon_0)$. To account for x-directed electric conductivity we need to connect to our TLM network a conductance given by

$$G_x = \sigma_e \frac{\Delta z \Delta y}{\Delta x} = \sigma_e \Delta\ell \qquad (6.20)$$

It is clear from the definition of σ_e that it generally depends on frequency so its introduction into the time-domain code can be done simply only if it is approximated by a single representative value. Special techniques will be presented in Section 6.4 to deal more accurately with frequency-dependent features. Magnetic losses may be introduced by incorporating a resistance into the TLM code as described in connection with (6.20),

$$R_x = \sigma_m \frac{\Delta z \Delta y}{\Delta x} = \sigma_m \Delta\ell \qquad (6.21)$$

where σ_m is the magnetic resistivity (Ωm^{-1}). Naturally, the scattering equations need to be derived again in the presence of stubs. Rather than doing this using the classical approach, we show in the next section another, more efficient, way of implementing scattering in a 3D TLM node.

6.2   A SIMPLE AND ELEGANT SCATTERING PROCEDURE IN 3D TLM

In the previous section we have shown how scattering may be implemented in 3D TLM. We now show how this process can be further streamlined and made more efficient. Our discussion resumes from the paragraph in the previous section introducing the four principles on which scattering is based. We have already mentioned in connection with Fig. 6.2 that there is no

DC connectivity at the center of the SCN, but it may be possible to derive equivalent voltages and currents. There is some algebraic effort involved in developing this technique, but it is worthwhile as it gives a very compact formulation for scattering for the most general node [34, 35]. I will illustrate the development by examining in detail conditions at the center of the node for the y-component of the electric field and the z-component of the magnetic field. Conditions for E_y are shown in Fig. 6.3(a). In addition to the y-polarized link lines I have introduced a capacitive stub (admittance Y_{oy}) and electric losses (conductance G_y). An average

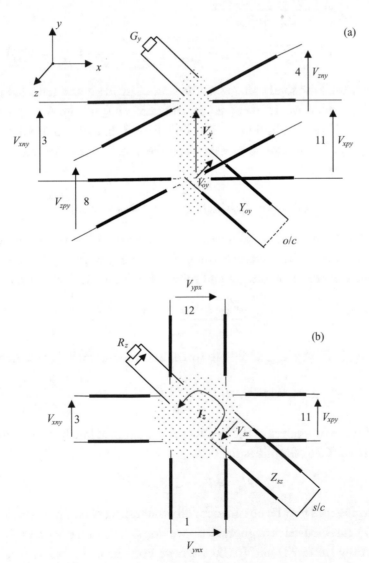

FIGURE 6.3: Same as in Fig. 6.2, but with open short-circuited stubs and electric/magnetic losses

"equivalent" voltage V_y may be obtained for the x-directed y-polarized line by imposing charge balance.

$$\left(V_{xny} Y_{xny} + V_{xpy} Y_{xpy}\right) \frac{\Delta t}{2} = V_y \left(Y_{xny} + Y_{xpy}\right) \frac{\Delta t}{2} \tag{6.22}$$

In (6.22) the LHS is the total charge on the two half lines and the RHS is the same charge on the two lines assuming an average voltage. In terms of the line impedances the equivalent voltage is obtained directly from this equation as

$$
\begin{aligned}
V_y &= \frac{V_{xny} Z_{xpy} + V_{xpy} Z_{xny}}{Z_{xpy} + Z_{xny}} \\
&= \frac{Z_{xpy}}{Z_{xny} + Z_{xpy}} \left(V_{xny}^i + V_{xny}^r\right) + \frac{Z_{xny}}{Z_{xny} + Z_{xpy}} \left(V_{xpy}^i + V_{xpy}^r\right)
\end{aligned}
\tag{6.23}
$$

where I have expressed the total voltages as the sum of incident and reflected pulses.

In a similar fashion we define an equivalent current I_z in Fig. 6.3(b) for the x-directed y-polarized line, where in addition to the link lines I have introduced an inductive stub (impedance Z_{sz}) and also magnetic losses (resistance R_z). The average current is obtained by imposing flux balance

$$\left(I_{xpy} Z_{xpy} - I_{xny} Z_{xny}\right) \frac{\Delta t}{2} = I_z \left(Z_{xny} + Z_{xpy}\right) \frac{\Delta t}{2} \tag{6.24}$$

where on the LHS we have the total magnetic flux linked with this line and on the RHS the same flux associated with the average current. Substituting in (6.24) for the total currents in terms of incident and reflected voltages and solving for the equivalent current we get

$$I_z = \frac{V_{xpy}^i - V_{xpy}^r}{Z_{xny} + Z_{xpy}} - \frac{V_{xny}^i - V_{xny}^r}{Z_{xny} + Z_{xpy}} \tag{6.25}$$

Multiplying (6.25) by Z_{xny} adding to (6.24) and solving for V_{xny}^r we obtain

$$V_{xny}^r = V_y + I_z Z_{xny} + V_{xny}^i \frac{Z_{xny} - Z_{xpy}}{Z_{xny} + Z_{xpy}} - V_{xpy}^i \frac{2 Z_{xny}}{Z_{xny} + Z_{xpy}} \tag{6.26}$$

Note that if the two halves of the x-directed y-polarized line have the same characteristic impedance Z_{xy} then (6.26) simplifies to

$$V_{xny}^r = V_y + I_z Z_{xy} - V_{xpy}^i \tag{6.27}$$

Similar expressions may be obtained for all other reflected components. However, neither (6.26) nor (6.27) can be used straightaway as we do not yet know V_y and I_z as a function of incident pulses only [in (6.23) and (6.25) we have both incident and reflected pulses on the RHS]. In order to get these equivalent quantities in terms of incident pulses only, we need to take account of the whole node shown in Fig. 6.3—not just the x-directed y-polarized lines. To find V_y [Fig. 6.3(a)] we impose charge conservation and charge balance on x-directed y-polarized lines and z-directed y-polarized lines.

Charge conservation (KCL) gives

$$
Y_{xny}\left(V_{xny}^i - V_{xny}^r\right) + Y_{xpy}\left(V_{xpy}^i - V_{xpy}^r\right) + Y_{zny}\left(V_{zny}^i - V_{zny}^r\right) + Y_{zpy}\left(V_{zpy}^i - V_{zpy}^r\right)
$$
$$
+ Y_{oy}\left(V_{oy}^i - V_{oy}^r\right) - G_y V_y = 0 \tag{6.28}
$$

The last term in (6.28) represents the charge reflected into the conductance representing losses—there is no incident pulse (not a storage component).

Using (6.23), charge balance for the x-directed y-polarized lines gives

$$
Y_{xny}V_{xny}^r + Y_{xpy}V_{xpy}^r = V_y\left(Y_{xny} + Y_{xpy}\right) - Y_{xny}V_{xny}^i - Y_{xpy}V_{xpy}^i \tag{6.29}
$$

A similar expression is obtained by imposing charge balance on the z-directed y-polarized lines.

$$
Y_{zny}V_{zny}^r + Y_{zpy}V_{zpy}^r = V_y\left(Y_{zny} + Y_{zpy}\right) - Y_{zny}V_{zny}^i - Y_{zpy}V_{zpy}^i \tag{6.30}
$$

Substituting (6.29) and (6.30) into (6.28) and solving for V_y we obtain

$$
V_y = 2\frac{Y_{xny}V_{xny}^i + Y_{xpy}V_{xpy}^i + Y_{zny}V_{zny}^i + Y_{zpy}V_{zpy}^i + Y_{oy}V_{oy}^i}{Y_{xny} + Y_{xpy} + Y_{zny} + Y_{zpy} + G_y} \tag{6.31}
$$

We now have the equivalent voltage as desired in terms of incident pulses only.

We now focus on the calculation of the equivalent current. Magnetic flux conservation (KVL) in Fig. 6.3(b) gives

$$
\left(V_{xpy}^i + V_{xpy}^r\right) - \left(V_{xny}^i + V_{xny}^r\right) + \left(V_{ynx}^i + V_{ynx}^r\right) - \left(V_{ypx}^i + V_{ypx}^r\right) - \left(V_{sz}^i + V_{sz}^r\right) - I_z R_z = 0
$$
$$
\tag{6.32}
$$

Using (6.24), flux balance for the x-directed y-polarized line gives

$$
V_{xpy}^r - V_{xny}^r = V_{xpy}^i - V_{xny}^i - I_z\left(Z_{xny} + Z_{xpy}\right) \tag{6.33}
$$

Similarly, flux balance for the y-directed x-polarized line gives

$$
V_{ypx}^r - V_{ynx}^r = V_{ypx}^i - V_{ynx}^i + I_z\left(Z_{ynx} + Z_{ypx}\right) \tag{6.34}
$$

Substituting (6.33) and (6.34) into (6.32) and solving for I_z we obtain

$$
I_z = 2\frac{V_{xpy}^i - V_{xny}^i + V_{ynx}^i - V_{ypx}^i - V_{sz}^i}{Z_{xny} + Z_{xpy} + Z_{ynx} + Z_{ypx} + Z_{sz} + R_z} \tag{6.35}
$$

This is as desired in terms of incident pulses only. We have now all we need to implement scattering—with (6.35) for I_z and (6.30) for V_y substituted into (6.26) or (6.27) for the simpler case we can obtain the reflected voltage V_{xny}^r. Similar expressions can be derived for the remaining reflected voltage pulses. In compact form, the scattering equations for the most general node are

$$V_{inj}^r = V_j \pm I_k Z_{inj} - V_{ipj}^i + h_{ij} \tag{6.36}$$
$$V_{ipj}^r = V_j \mp I_k Z_{ipj} - V_{inj}^i + h_{ij} \tag{6.37}$$

where the upper signs apply for $(i, j, k) \in \{(x, y, z), (y, z, x), (z, x, y)\}$ and the lower signs for $(i, j, k) \in \{(x, z, y), (y, x, z), (z, y, x)\}$. The equivalent voltages and currents are given by

$$V_i = 2 \frac{Y_{kni} V_{kni}^i + Y_{kpi} V_{kpi}^i + Y_{jni} V_{jni}^i + Y_{jpi} V_{jpi}^i + Y_{oi} V_{oi}^i}{Y_{kni} + Y_{kpi} + Y_{jni} + Y_{jpi} + Y_{oi} + G_i} \tag{6.38}$$

$$I_i = 2 \frac{V_{jpk}^i - V_{jnk}^i + V_{knj}^i - V_{kpj}^i - V_{si}^i}{Z_{jnk} + Z_{jpk} + Z_{knj} + Z_{kpj} + Z_{si} + R_i} \tag{6.39}$$

where $(i, j, k) \in \{(x, y, z), (y, z, x), (z, x, y)\}$. The h-factors in (6.36) and (6.37) are given by

$$h_{ij} = \frac{Z_{inj} - Z_{ipj}}{Z_{inj} + Z_{ipj}} \left(V_{inj}^i - V_{ipj}^i \right) \tag{6.40}$$

Note that for the common case of the same impedance for both halves of lines with the same direction and polarization h-factors are equal to zero. The reflected pulses into the stubs (open circuit and short circuit) are

$$V_{oi}^r = V_i - V_{oi}^i \tag{6.41}$$
$$V_{si}^r = I_i Z_{si} + V_{si}^i \tag{6.42}$$

where $i \in \{x, y, z\}$. The voltages reflected into the components representing electric and magnetic losses in the i-polarization are V_i and $I_i R_i$, respectively.

Thus, the scattering procedure is to calculate first the three equivalent currents and the three equivalent voltages from (6.39) and (6.38), then the h-factors from (6.40) (if non-zero) and after that the reflected pulses into the link lines (6.36) and (6.37), the stubs (6.41) and (6.42).

Scattering coefficients as components of the matrix in (6.8) may also be obtained, if required (e.g. for dispersion analysis), for all nodes and are available in the literature [4, 34]. Most TLM nodes in common use are not as general as assumed in this section therefore scattering is further simplified.

6.3   PARAMETER CALCULATION AND CLASSIFICATION OF TLM NODES

We have obtained the scattering matrix in (6.8) for a cubic node where all link lines have the same characteristic impedance and another scattering procedure in the last section for a General SCN (GSCN) node with stubs and losses and also potentially all half lines having a different characteristic impedance. You may wonder under what circumstances this GSCN may be employed and how all these parameters may be calculated. We will see that many different choices are possible resulting in a number of different nodes to model fields but the numerical properties (dispersion, number of arithmetic operations, storage) differ. Firstly, we focus on the case where the impedance of an i-directed j-polarized link line is the same on both sides of the node, i.e., $Z_{inj} = Z_{ipj} = Z_{ij}$. Nodes where this equality does not hold are rare and only used to obtain unequal arm lengths and adjust better near boundaries [34, 36]. For obtaining the parameters of any node we need to enforce the following conditions:

- The total capacitance in each of the three coordinate directions must be equal to the capacitance of the block of space represented by the cell.

- The total inductance in each of the three coordinate directions must be equal to the inductance of the block of space represented by the cell.

- Synchronism must be maintained (same propagation time on all link lines).

We have 18 unknowns: 6 link line capacitances, 6 link line inductances, 3 stub capacitances, and 3 stub inductances. The equation we obtain from the first of the above conditions is

$$C_{yx}\Delta y + C_{zx}\Delta z + C_{ox} = \varepsilon_x \frac{\Delta y \Delta z}{\Delta x} \qquad (6.43)$$

The first term in this equation is the capacitance in the x-polarization of the y-directed line, the second term is for the z-directed line, and the third the contribution of the x-polarized stub. The RHS is the cell capacitance assuming that the permittivity in the x-direction is ε_x. Similar expressions may be written for the other two polarizations

$$C_{zy}\Delta z + C_{xy}\Delta x + C_{oy} = \varepsilon_y \frac{\Delta x \Delta z}{\Delta y} \qquad (6.44)$$

$$C_{xz}\Delta x + C_{yz}\Delta y + C_{oz} = \varepsilon_z \frac{\Delta y \Delta x}{\Delta z} \qquad (6.45)$$

We now impose the second condition

$$L_{yz}\Delta y + L_{zy}\Delta z + L_{sx} = \mu_x \frac{\Delta y \Delta z}{\Delta x} \qquad (6.46)$$

The first term in this equation is the y-directed z-polarized link line inductance, the second is for the z-directed y-polarized link line and the third for the contribution of the x-directed inductive stub. All three terms contribute to the x-directed magnetic field. The RHS is the inductance of the cell associated with the x-component of the magnetic field. Similar expressions are obtained for the remaining two components:

$$L_{zx}\Delta z + L_{xz}\Delta x + L_{sy} = \mu_y \frac{\Delta x \Delta z}{\Delta y} \tag{6.47}$$

$$L_{xy}\Delta x + L_{yx}\Delta y + L_{sz} = \mu_z \frac{\Delta y \Delta x}{\Delta z} \tag{6.48}$$

The delay time along a line is given by (3.8) which applies to each of the six lines and taking into account that we use per unit length quantities gives

$$\begin{aligned} \Delta t = \Delta x \sqrt{C_{xy} L_{xy}} \qquad \Delta t = \Delta x \sqrt{C_{xz} L_{xz}} \qquad \Delta t = \Delta y \sqrt{C_{yz} L_{yz}} \\ \Delta t = \Delta y \sqrt{C_{yx} L_{yx}} \qquad \Delta t = \Delta z \sqrt{C_{zx} L_{zx}} \qquad \Delta t = \Delta z \sqrt{C_{zy} L_{zy}} \end{aligned} \tag{6.49}$$

Equations (6.43)–(6.49) are 12 in total and we have 18 unknowns. We have thus six degrees of freedom and we can derive many different configurations by imposing additional constraints. This is the origin of the different nodes. It is not appropriate in this text to go into the details of each node but a general classification will be given to assist the reader in making sense of the various options.

We can put these equations into a compact form by using the synchronism equation to express all inductances and capacitances in terms of the characteristic impedances and admittances

$$\begin{aligned} Z_{ij} = \frac{L_{ij}\Delta i}{\Delta t} \qquad Y_{ij} = \frac{C_{ij}\Delta i}{\Delta t} \\ Z_{sk} = \frac{2 L_{sk}}{\Delta t} \qquad Y_{ok} = \frac{2 C_{0k}}{\Delta t} \end{aligned} \tag{6.50}$$

Then Eqs. (6.43)–(6.48) reduce to

$$Y_{ik} + Y_{jk} + \frac{Y_{ok}}{2} = \varepsilon_k \frac{\Delta i \Delta j}{\Delta k \Delta t} \tag{6.51}$$

$$Z_{ij} + Z_{ji} + \frac{Z_{sk}}{2} = \mu_k \frac{\Delta i \Delta j}{\Delta k \Delta t} \tag{6.52}$$

If we take advantage of the six degrees of freedom to impose the condition that all six link line impedances are equal to Z_0, the intrinsic impedance of the background medium (free space in most cases, $Z_0 = \sqrt{\mu_0/\varepsilon_0}$), we obtain the *stub-loaded SCN* [4, 32]. Equations (6.51) and (6.52) become

$$2Y_0 + \frac{Y_{ok}}{2} = \varepsilon_k \frac{\Delta i \, \Delta j}{\Delta k \, \Delta t} \qquad (6.53)$$

$$2Z_0 + \frac{Z_{sk}}{2} = \mu_k \frac{\Delta i \, \Delta j}{\Delta k \, \Delta t} \qquad (6.54)$$

The open-circuit stub admittances and short-circuit stub impedances are then obtained from

$$Y_{ok} = 2Y_0 \left(\frac{\varepsilon_{rk}}{c \, \Delta t} \frac{\Delta i \, \Delta j}{\Delta k} - 2 \right) \qquad (6.55)$$

$$Z_{sk} = 2Z_0 \left(\frac{\mu_{rk}}{c \, \Delta t} \frac{\Delta i \, \Delta j}{\Delta k} - 2 \right) \qquad (6.56)$$

where $\varepsilon_{rk}, \mu_{rk}$ are the relative permittivity and permeability in the k-direction, respectively. Clearly, for stability, the time step must be chosen such that the expressions in the brackets in (6.55) and (6.56) are positive. This imposes a maximum permissible value for Δt. In free space and for a cubic node we see that the maximum permissible time step is as expected from (6.12). However, for a noncubic node the maximum time step is affected by the smallest nodal dimension and also the aspect ratio of the node [34]. Indiscriminate "grading" of the cell to introduce long and thin shapes may result in unacceptably small time steps.

As expected the node described above is not the only choice. We may develop a Type 1 Hybrid SCN (HSCN) where all inductance is modeled by the link lines and hence we do not need inductive stubs. This implies that the impedance of link lines modeling different magnetic field components may be different. Details of this node may be found in [37, 38]. Alternatively, a Type 2 HSCN may be developed where all capacitance is modelled by the link lines (no capacitive stubs) [39]. Similar constraints apply as for the stub-loaded SCN as regards the maximum permissible time step. The type 1 HSCN is the node most often used in commercial TLM codes. There are several other nodes described in the literature, e.g., MSCN [40], ASCN [41]. I also mention nodes that are based on non-Cartesian meshes [25]. There are also nodes suitable for frequency-domain implementations of TLM [42, 43]; these publications can be consulted if required.

6.4   FIELD OUTPUT IN 3D TLM

Electric/magnetic fields, current/charge densities are obtained in a similar way as for the 2D node. As regards fields there are two possibilities: mapping at the cell center or at the cell boundaries.

First we may define fields at the center of the node from the equivalences

$$E_i = -\frac{V_i}{\Delta i} \tag{6.57}$$

$$H_i = \frac{I_i}{\Delta i} \tag{6.58}$$

where the equivalent voltages and currents are obtained from (6.38) and (6.39). As an example for the 12-port SCN (no stubs) in free space the x-component of the electric field is

$$E_x = -\frac{V_x}{\Delta x} = -\frac{V^i_{ynx} + V^i_{ypx} + V^i_{znx} + V^i_{zpx}}{2\Delta\ell} = -\frac{V_1 + V_{12} + V_2 + V_9}{2\Delta\ell} \tag{6.59}$$

where both port notations have been used. Similarly, the x-component of the magnetic field is

$$H_x = \frac{I_x}{\Delta x} = \frac{V^i_{zny} + V^i_{ypz} - V^i_{ynz} - V^i_{zpy}}{2Z_0\Delta\ell} = \frac{V_4 + V_7 - V_5 - V_8}{2Z_0\Delta\ell} \tag{6.60}$$

Similar expressions apply for the remaining field components. For the stub-loaded SCN we have 18 voltages pulses (12 from the link lines and three each from the capacitive and inductive stubs) that map to six field components (three for electric field and three for magnetic field) at the node center. Therefore there is no one-to-one correspondence. A one-to-one (bijective) mapping may be established at the cell boundaries. With reference to Fig. 6.4 we are interested in the fields at the RHS of the boundary between the two cells. We illustrate as an example conditions on the x-directed y-polarized line. The total y-polarized voltage on the RHS of the boundary at time $k + 1/2$ is made out of the sum of the voltage reflected from the node on the right at time k and the voltage incident on the node at time $k + 1$ (this originates from the pulse reflected from the node on the left at time k). The two pulses meet at time $k + 1/2$.

$$_{k+1/2}V_y = {_k}V^r_{xny} + {_{k+1}}V^i_{xny} \tag{6.61}$$

Similarly, the current there is

$$_{k+1/2}I_y = \frac{_k V^r_{xny} - {_{k+1}}V^i_{xny}}{Z_{xny}} \tag{6.62}$$

Thus the two pulses $_k V^r_{xny}$, $_{k+1}V^i_{xny}$ at the boundary of the cell to the right map to V_y and I_z, i.e., E_y and H_z. Similar expressions may be derived for all other field components tangential to the cell boundaries.

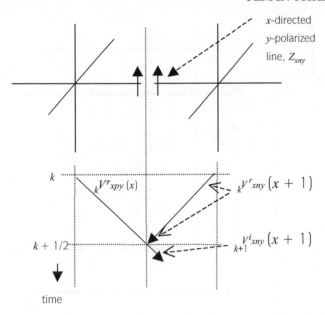

FIGURE 6.4: Schematic showing the propagation of pulses used for field calculation at cell boundaries

6.5 MODELING OF GENERAL MATERIAL PROPERTIES IN TLM

In the previous section we mentioned briefly how losses may be introduced in TLM models through conductance and resistance for electric and magnetic losses, respectively. However, it is evident that in this way only a fixed loss value may be introduced which is independent of the frequency. Many materials exhibit strong loss dependence on frequency and in addition anisotropy, chirality, and nonlinear behavior. In order to account for the entire range of material properties more sophisticated material models need to be developed that go beyond the mere introduction of fixed loss components. The topic is a complex one but of great practical value. It cannot be fully treated in the present text. I will give, however, a introduction to the relevant techniques so that the reader may find it easier to access more advanced work in the literature [44–50].

I will illustrate the modeling philosophy by focusing on modeling *frequency-dependent conductivity*. One can view a frequency-dependent component as introducing a transfer function (TF) $H(\omega)$. Modeling this TF in a frequency-domain model is straightforward. However, in a TD model, where frequency does not appear as a variable, the TF cannot be accounted for without first converting to the discrete time-domain. The problem posed is similar to the one in electrical filter theory: given an analog filter (specified in the FD) derive an equivalent digital filter (specified in the discrete TD). There are formal methods in filter theory to do just this,

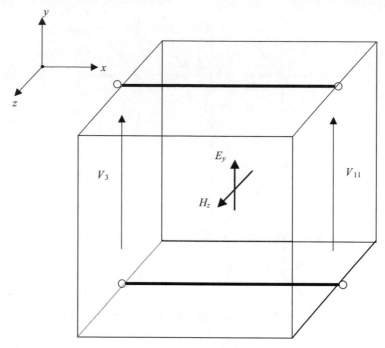

FIGURE 6.5: The TLM node for the study of 1D propagation in a general medium

e.g., through the application of z-transforms and we have to do something similar. Since the algorithm resulting on the completion of this process is in the discrete TD it can be directly embedded into the TLM algorithm. To reduce complexity and thus illustrate better the essence of the approach I develop fully a 1D model that can be readily generalized for 3D modeling [47–50]. The network configuration is shown schematically in Fig. 6.5. We need to derive the 1D field equation and the corresponding network equation, establish equivalence and then solve the network in the discrete TD. The process is similar to that described in Section 2.2 but with the important difference that now the electrical conductivity is a function of frequency. For σ independent of frequency, the electric current density and the electric filed are related simply by $J_e = \sigma E$. Frequency dependence of a quantity implies delays and that the time history is important. A simple multiplication in the frequency-domain is translated into a convolution in the time-domain.

$$J_e(x, t) = \int_0^t \sigma_e(x, \tau) E(x, t - \tau) d\tau = \sigma_e(x, t) * E(x, t) \qquad (6.63)$$

With this complication, Faraday's laws and Ampere's laws in 1D reduce to

$$-\frac{\partial H_z(x, t)}{\partial x} = J_{efy}(x, t) + \sigma_e(x, t) * E_y(x, t) + \varepsilon_0 \frac{\partial E_y(x, t)}{\partial t} \qquad (6.64)$$

$$-\frac{\partial E_y(x, t)}{\partial x} = \mu_0 \frac{\partial H_z(x, t)}{\partial t} \qquad (6.65)$$

In (6.64) the term J_{efy} is to account for the possible presence of free electric sources. We now seek to derive a circuit analog to these equations. Unlike the work we have done so far, where we have used dispersionless TLs to construct nodes, we must now recognize that the TLs may contain components that in addition to normal spatial dependence also exhibit time dependence. In the particular case of the field problem described by Eqs. (6.64) and (6.65) we anticipate that resistive losses which depend on (x, t) must be incorporated. We start with a TL equivalent as shown in Fig. 6.6. In addition to the normal L–C components we have a shunt branch consisting of a resistance $G_e(x, t)$ to represent the losses and a current source $I_{fy}(x, t)$ to account for the possible presence of free sources. Applying KCL in this circuit gives

$$-\frac{\partial I_z(x, t)}{\partial x} \Delta\ell = I_3 + I_{11} = I_{fy}(x, t) + G_e(x, t) * V_y(x, t) + C \frac{\partial V_y(x, t)}{\partial t} \qquad (6.66)$$

I will now demonstrate that (6.66) is analogous to (6.64). Dividing both sides of (6.66) by $\Delta\ell$ and setting $G_e(x, t)/\Delta\ell = \sigma_e(x, t)$ the medium electrical conductivity, and $C/\Delta\ell = \varepsilon_0$, we get

$$-\frac{\partial I_z(x, t)}{\partial x} = \frac{I_{fy}(x, t)}{\Delta\ell} + \sigma_e(x, t) * V_y(x, t) + \varepsilon_0 \frac{\partial V_y(x, t)}{\partial t} \qquad (6.67)$$

We now define the following analogies between field and circuit parameters

$$E_y(x, t) = -\frac{V_y(x, t)}{\Delta\ell}, \qquad H_z(x, t) = -\frac{I_z(x, t)}{\Delta\ell} \qquad (6.68)$$

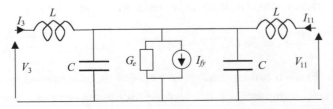

FIGURE 6.6: Parameters of the line representing a medium with frequency-dependent electrical conductivity

and the free current density as

$$J_{efy}(x, t) = -\frac{I_{fy}(x, t)}{(\Delta\ell)^2} \qquad (6.69)$$

Substituting (6.68) and (6.69) into (6.67) gives an equation identical to (6.64) thus showing that solution of the circuit in Fig. 6.6 will give us complete information about the corresponding field problem. We can also apply KVL and show in a similar way that we obtain the circuit analog of (6.65). Let us now focus on solving (6.67). To do this we first go through a process of normalization to simplify the various terms in preparation for solution. We introduce normalized time T and space X variables by expressing the relevant derivatives as

$$\frac{\partial}{\partial x} = \frac{1}{\Delta\ell}\frac{\partial}{\partial X}, \qquad \frac{\partial}{\partial t} = \frac{1}{\Delta t}\frac{\partial}{\partial T} \qquad (6.70)$$

We also represent the current by a new quantity i_z, which has the dimensions of volts, defined by the following expression

$$-\frac{I_z(X, T)}{\Delta\ell} = -\frac{i_z(X, T)}{\Delta\ell}\frac{1}{\eta_0} \qquad (6.71)$$

Substituting (6.70) and (6.71) into (6.67) and multiplying both sides by $\Delta\ell\eta_0$ we get

$$-\frac{\partial i_z(X, T)}{\partial X} = \frac{I_{fy}(X, T)}{\Delta\ell}\Delta\ell\eta_0 + \sigma_e(X, T) * V_y(X, T)\Delta\ell\eta_0 + \eta_0\varepsilon_0\frac{\Delta\ell}{\Delta t}\frac{\partial V_y(X, T)}{\partial T}$$

The first term on the RHS of this equation is represented by a quantity i_{fy} (dimensions of volts) defined as

$$i_{fy}(X, T) = -J_{efy}(X, T)(\Delta\ell)^2 \eta_0 \qquad (6.72)$$

The second term is equal to $g_e(X, T) * V_y(X, T)$ where we have defined a normalized conductivity as

$$g_e(X, T) = \sigma_e(X, T)\Delta\ell\eta_0 \qquad (6.73)$$

Recognizing that $\Delta\ell/\Delta t = c$ results in the coefficient of the derivative in the third term being equal to one. Hence, the circuit equation reduces to the following form

$$-\frac{\partial i_z(X, T)}{\partial X} = i_{fy}(X, T) + g_e(X, T) * V_y(X, T) + \frac{\partial V_y(X, T)}{\partial T} \qquad (6.74)$$

The circuit in Fig. 6.6 can be redrawn in terms of the normalized quantities in (6.74) by observing that the $L–C$ components at the two ends of the segment can be expressed as two TL segments of characteristic impedances η_0 and since we have normalized all other impedances to this value the normalized impedance of each TL is equal to one. This is shown in Fig. 6.7(a)

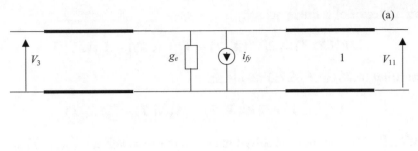

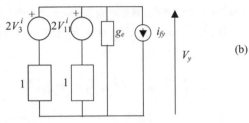

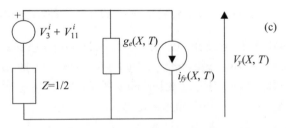

FIGURE 6.7: TL representation in normalized form of the problem in Fig. 6.6 (a), and Thevenin representations (b) and (c)

with the corresponding Thevenin equivalent circuit in Fig. 6.7(b). Replacing the two parallel branches with voltage sources by their Thevenin equivalent we obtain the circuit shown in Fig. 6.7(c). Summing the three currents on the branches to zero (KCL)

$$\frac{V_y(X, T) - [V_3^i(X, T) + V_{11}^i(X, T)]}{1/2} + g_e(X, T) * V_y(X, T) + i_{fy}(X, T) = 0$$

and rearranging we obtain

$$2[V_3^i(X, T) + V_{11}^i(X, T)] - i_{fy}(X, T) = 2V_y(X, T) + g_e(X, T) * V_y(X, T) \qquad (6.75)$$

This is (6.74) in terms of the incident voltage pulses. The LHS of this equation is essentially the excitation of this segment at time T (incident pulses from the left and from the right and any free sources present within the segment) and it is convenient to express this "source term" in terms of a single quantity.

$$2V_y^r(X, T) = 2[V_3^i(X, T) + V_{11}^i(X, T)] - i_{fy}(X, T) \tag{6.76}$$

Substituting (6.76) into (6.75) we obtain

$$2V_y^r(X, T) = 2V_y(X, T) + g_e(X, T) * V_y(X, T) \tag{6.77}$$

Equation (6.77) must now be solved to obtain the total voltage $V_y(X, T)$ at this segment. Note that this is not straightforward, as the unknown appears also inside the convolution integral (second term on the RHS). Transforming this equation in the discrete time-domain using the z-operator gives

$$2V_y^r(X, z) = 2V_y(X, z) + g_e(X, z)V_y(X, z) \tag{6.78}$$

where in the z-domain the convolution becomes a simple multiplication. We manipulate this expression further by putting the frequency-dependent term $g_e(X, z)$ in the form

$$(1 + z^{-1})g_e(X, z) = g_{e0}(X) + z^{-1}[g_{e1}(X) + \bar{g}_e(X, z)] \tag{6.79}$$

What this does is to make the frequency-dependent part of the conductivity to depend on the field at the previous time step (delay operator z^{-1} in front of the brackets on the RHS). This gives an explicit algorithm. Multiplying (6.78) by $(1 + z^{-1})$, substituting from (6.79) and solving for $V_y(X, z)$ after some algebraic manipulation

$$V_y(X, z) = T_e\left[2V_y^r(X, z) + z^{-1}S_{ey}(X, z)\right] \tag{6.80}$$

where

$T_e = (2 + g_{e0})^{-1}$ is a frequency-independent term, and

$$S_{ey}(X, z) = 2V_y^r(X, z) + \kappa_e(X)V_y(X, z) - \bar{g}_e(X, z)V_y(X, z) \tag{6.81}$$

The frequency independent term $\kappa_e(X) = -(2 + g_{e1}(X))$. We note from (6.80) that the total voltage at any one instant depends on the current excitation (first term on the RHS) and contributions from the *previous* time step (second term on the RHS). The quantity $S_{ey}(X, z)$ which represents this stored information is called the main accumulator. It is given by (6.81) and includes information from the previous excitation, total voltage and also the recent history of the material [the third term in (6.81)]. Since we do not know at this stage the details of the material we cannot give substance to the third term so we introduce yet another accumulator (let us call it the conduction accumulator) S_{ec}, where

$$S_{ec}(X, z) = -\bar{g}_e(X, z)V_y(X, z) \tag{6.82}$$

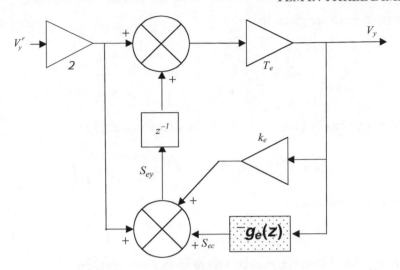

FIGURE 6.8: Flow diagram in the z-domain to implement scattering

Equations (6.80)–(6.82) are embodied into the graph shown in Fig 6.8 (essentially a digital filter algorithm). This is the formal solution to our problems of determining the total voltage V_y in terms of the excitation. No further progress can be made without an exact specification for the frequency-dependent term $-\bar{g}_e (X, z)$. Depending on the form of this function (and it may be a complicated function of z) we may end up needing to produce yet another filter to replace the box labeled $-\bar{g}_e (X, z)$ in Fig. 6.8. Also note that if the frequency dependence for the material is in the conductivity only the scheme in Fig. 6.8 will work for us—we only need to work out in each case the details of this box. I will do this for one particular material, unmagnetized plasma [47], where the frequency-dependent conductivity has the form

$$g_e(s) = \frac{g_{ec}}{1 + s\tau_c} \qquad (6.83)$$

For simplicity, I suppress hereafter showing space dependence. In (6.83) s is the Laplace variable. We transform the conductivity into the z-domain by applying the bilinear transformation [18].

$$s \rightarrow \frac{2}{\Delta t}\frac{1 - z^{-1}}{1 + z^{-1}} \qquad (6.84)$$

Substituting for s into (6.83) we obtain after some algebra

$$g_e (z) = \frac{K_1 \left(1 + z^{-1}\right)}{1 - z^{-1}a_1} \qquad (6.85)$$

where the two constants are given by

$$K_1 = \frac{g_{ec}}{\frac{2\tau_c}{\Delta t} + 1}, \qquad a_1 = \frac{\frac{2\tau_c}{\Delta t} - 1}{\frac{2\tau_c}{\Delta t} + 1} \qquad (6.86)$$

Multiplying (6.85) by $\left(1 + z^{-1}\right)$ we obtain after some algebra

$$\left(1 + z^{-1}\right) g_e \left(z\right) = K_1 + z^{-1} \left(\frac{b_0 + z^{-1}b_1}{1 - z^{-1}a_1}\right) \qquad (6.87)$$

where the two new constants are

$$b_0 = K_1 \left(2 + a_1\right), \qquad b_1 = K_1 \qquad (6.88)$$

Comparing (6.87) with the standard form in (6.79) we obtain

$$g_{e0} = K_1, \qquad g_{e1} = 0, \qquad \bar{g}_e \left(z\right) = \frac{b_0 + z^{-1}b_1}{1 - z^{-1}a_1}, \qquad \kappa_e = -2 \qquad (6.89)$$

We now have the functional form of the material properties for the magnetized plasma and we can thus proceed to design an algorithm to calculate the conduction accumulator in (6.82).

$$S_{ec} \left(z\right) = -\frac{b_0 + z^{-1}b_1}{1 - z^{-1}a_1} V_y \left(z\right) \qquad (6.90)$$

We define a state variable $X_0(z)$ such that

$$S_{ec} \left(z\right) = -(b_0 + z^{-1}b_1)X_0 \left(z\right) \qquad (6.91)$$

Substituting (6.91) into (6.90) we obtain

$$X_0 \left(z\right) = z^{-1}a_1 X_0 \left(z\right) + V_y \left(z\right) \qquad (6.92)$$

This equation allows us to calculate the state variable at the current time step as a function of its value at the previous time step and the current value of the total voltage. Equations (6.91) and (6.92) may be implemented by the flow graph shown in Fig. 6.9. We can confirm that the transfer function of this filter is as desired by checking directly or by applying Mason's rule

$$\text{TF} = \frac{F_1 + F_2}{1 - L_1} \qquad (6.93)$$

where F_i and L_i are the forward and loop gains. For the scheme in Fig. 6.9 these gains are $F_1 = -b_0$, $F_2 = -z^{-1}b_1$, $L_1 = z^{-1}a_1$. Substituting these gains into (6.93) we obtain (6.90) as

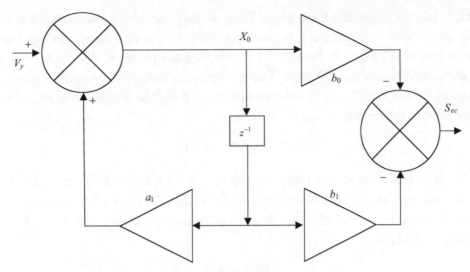

FIGURE 6.9: Structure of the shaded box in Fig. 6.8 for the particular example of an unmagnetized plasma

expected. Figure 6.9 replaces box $-\bar{g}_e\,(z)$ in Fig. 6.8. There is one delay element in Figs. 6.8 and 6.9 introduces another one, hence two quantities need to be remembered (stored) to advance the calculation. In addition the incident voltages from ports 3 and 11 are required.

The complete algorithm for this particular material is as follows:

- Obtain excitation at the current time step from (6.76).
- Calculate the total voltage $V_y(X,\,T)$ at the current time step from (6.80) using the current value of the excitation and the main accumulator value obtained from the previous time step.
- Calculate the current value of the state variable using its previous value and the current value of the total voltage using (6.92).
- Calculate the current value of the conduction accumulator from (6.91).
- Calculate the current value of the main accumulator from (6.81).
- Calculate the reflected voltages from

$$V_3^r(X,\,T) = V_y(X,\,T) - V_3^i(X,\,T), \quad V_{11}^r(X,\,T) = V_y(X,\,T) - V_{11}^i(X,\,T) \quad (6.94)$$

- Implement connections to the adjacent segments and repeat the process given above for the next time step.

How may we generalize these ideas? First, we may consider introducing a full model of the magnetic properties that requires the incorporation of series components to supplement the shunt components in Fig. 6.6. Second, we may extend the model to 3D to take full account of the spatial structure of the 3D node. The principles for doing this work are similar to those explained so far but the algebra is more complex and thus it is not included here. In general terms we need to update Eq. (6.76) to the form

$$F^r = R_1^T V^i - 0.5 V_f \tag{6.95}$$

where F^r represents the node excitation consisting of the incident voltage pulses (first term on the RHS), and any free sources (second term on the RHS). The matrix R_1^T represents the TLM process and the vector V_f includes electric and magnetic sources. Equation (6.78) can now be put in the general form

$$F = t(z) F^r \tag{6.96}$$

where F is the vector representing electric and magnetic fields and $t(z)$ is a matrix representing the transmission properties of the material. In the presence of series components, the two equations in (6.94) must be modified to account for the total voltage at each port that now depend on the shunt voltages (V_y in the 1D formulation for electric materials) and also terms due to the series components. The calculation of the reflected components may then be put in the general form

$$V^r = RF - PV^i \tag{6.97}$$

where the matrix R represents the reflection process in the TLM node and P is used to reorder the incident voltage pulses as required.

A graph representing the TLM scattering process in the most general case is shown in Fig. 6.10. We note that the material properties are represented by the matrix $t(z)$ which can be adapted to whichever material that is appropriate for this cell. The other three matrices are representing the spatial properties of the node. Using this graph and the incident and source vectors we may calculate the reflected vector. Full details of this process and examples of particular materials (box $[t(z)]$) may be found in [47–50].

6.6   MODELING THIN COMPLEX PANELS IN 3D TLM

In many engineering applications where an electromagnetic design is required, it is necessary to develop and implement models of thin panels that may incorporate a range of fine features. An example is a thin metal panel, part of an equipment cabinet, with a large number of small ventilation holes. It is well-known that the EM shielding effectiveness of this arrangement is

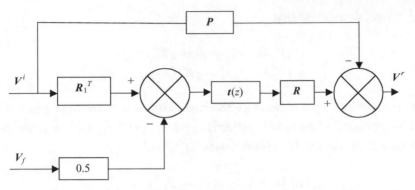

FIGURE 6.10: Flow diagram for scattering in a 3D node representing a cell consisting of a general dispersive material

better than a panel with a single large opening of an area equal to the total area of the small holes [51]. For this reason, such configurations are very common. The modeler is therefore confronted with the difficulty of modeling a thin panel and a large number of features with dimensions much smaller than the mesh resolution. This is another example of a multiscale problem similar in nature to the problem of embedding a thin wire in a coarse mesh that we have addressed in Section 4.6. As always, one may seek refuge in a larger, faster computer and increase resolution to cope with the very fine detail. Apart from the huge computational cost of this option the analyst needs to input into the mesh the details of every single hole—a task which is extremely time consuming. For both these reasons we seek other more efficient ways to implement such features into a mesh. I will illustrate the technique with reference to a thin perfectly conducting panel with perforations as shown in Fig. 6.11. We embed the panel between nodes so it influences the TLM connection process. In the normal way, in the absence

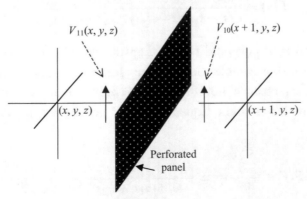

FIGURE 6.11: Connection between adjacent nodes in the presence of a thin frequency selective panel

of the panel connection means that

$$
\begin{aligned}
{}_{k+1}V_{10}^{i}(x+1,y,z) &= {}_{k}V_{11}^{r}(x,y,z) \\
{}_{k+1}V_{11}^{i}(x,y,z) &= {}_{k}V_{10}^{r}(x+1,y,z)
\end{aligned}
\tag{6.98}
$$

In the presence of a panel between the two nodes the connection process indicated by (6.98) must be modified. If we have a perfectly conducting solid panel matters are straightforward. We expect each pulse to be reflected with an opposite sign.

$$
\begin{aligned}
{}_{k+1}V_{10}^{i}(x+1,y,z) &= -{}_{k}V_{10}^{r}(x+1,y,z) \\
{}_{k+1}V_{11}^{i}(x,y,z) &= -{}_{k}V_{11}^{r}(x,y,z)
\end{aligned}
\tag{6.99}
$$

Equations (6.98) and (6.99) represent the two extreme cases—total transparency and total reflection. The case depicted in Fig. 6.11 is an intermediate one—there will be some reflection and some transmission. Moreover, both reflection and transmission will be frequency dependent as the perforations are selective in the way they allow different frequencies to penetrate. We see therefore that the connection must incorporate in it scattering and that this scattering is frequency dependent. In a schematic form connection now is equivalent to

$$
\begin{bmatrix} {}_{k+1}V_{10}^{i}(x+1,y,z) \\ {}_{k+1}V_{11}^{i}(x,y,z) \end{bmatrix} = \begin{bmatrix} R & T \\ T & R \end{bmatrix} \begin{bmatrix} {}_{k}V_{10}^{r}(x+1,y,z) \\ {}_{k}V_{11}^{r}(x,y,z) \end{bmatrix}
\tag{6.100}
$$

where R and T are the frequency-dependent reflection and transmission coefficients. The challenge is to implement this frequency-dependent scattering in the time-domain. The procedure is not unlike the one we explored in the last section. We start by assuming a general functional form for the scattering parameters (R or T) given by

$$
F(s) = \frac{b_{NP}(s-s_{z0})(s-s_{z1})\dots(s-s_{z(NP-1)})}{(s-s_{p0})(s-s_{p1})\dots(s-s_{p(NP-1)})}
\tag{6.101}
$$

where NP is the number of poles and s is the Laplace variable. The zero frequency and pole frequency s_{zi}, s_{pi} are real or complex, in the later case in conjugate pairs. We postpone until later in this section the question of how to obtain this function for particular configurations and we focus first on its implementation. Equation (6.101) may be put in a Padé form

$$
F(s) = \frac{\sum_{i=0}^{NP} b_i s^i}{\sum_{i=0}^{NP} a_i s^i} = \frac{b_0 + b_1 s + b_2 s^2 + \dots + b_{NP}s^{NP}}{a_0 + a_1 s + a_2 s^2 + \dots + s^{NP}}
\tag{6.102}
$$

How can we implement a scattering such as described in (6.102) in the discrete time-domain? We proceed by applying the impulse invariant transformation [18].

$$s - s_{zi} \leftrightarrow \frac{-s_{zi}}{1 - \beta_{zi}}(1 - z^{-1}\beta_{zi}) \tag{6.103}$$

where $\beta_{zi} = e^{s_{zi}\Delta t}$. Substituting (6.103) into (6.102) we obtain

$$F(z) = \left\{ b_{\mathrm{NP}} \prod_{i=0}^{\mathrm{NP}-1} \frac{s_{zi}(1 - \beta_{pi})}{s_{pi}(1 - \beta_{zi})} \right\} \prod_{i=0}^{\mathrm{NP}-1} \frac{1 - z^{-1}\beta_{zi}}{1 - z^{-1}\beta_{pi}} = B_0 \prod_{i=0}^{\mathrm{NP}-1} \frac{1 - z^{-1}\beta_{zi}}{1 - z^{-1}\beta_{pi}} \tag{6.104}$$

where we have designated the z-independent quantity in the angled brackets by B_0. The first term in the numerator and denominator after we work out the products is 1, hence we can put (6.104) in the form

$$F(z) = \frac{B_0 + \sum\limits_{i=1}^{\mathrm{NP}} B_i z^{-i}}{1 + \sum\limits_{i=1}^{\mathrm{NP}} A_i z^{-i}} \tag{6.105}$$

Adding and subtracting the term $\sum\limits_{i=0}^{\mathrm{NP}} B_0 A_i z^{-i}$ in the numerator, designating $B_i' = B_i - B_0 A_i$ and simplifying we obtain

$$F(z) = B_0 + \frac{\sum\limits_{i=1}^{\mathrm{NP}} B_i' z^{-i}}{1 + \sum\limits_{i=1}^{\mathrm{NP}} A_i z^{-i}} \tag{6.106}$$

Typically, (6.106) represents a reflection coefficient relating V^r and V^i, i.e.,

$$V^r = B_0 V^i + \frac{V^i \sum\limits_{i=1}^{\mathrm{NP}} B_i' z^{-i}}{1 + \sum\limits_{i=1}^{\mathrm{NP}} A_i z^{-i}} \tag{6.107}$$

We define state variables X_i such that

$$V^r = B_0 V^i + \sum\limits_{i=1}^{\mathrm{NP}} B_i' X_i \tag{6.108}$$

The second term on the RHS of (6.107) and (6.108) are equal by definition and this equality may be put in a matrix form.

$$
V^i \begin{bmatrix} B'_1 & . & . & B'_{NP} \end{bmatrix} \begin{bmatrix} z^{-1} \\ . \\ . \\ z^{-NP} \end{bmatrix} = \begin{bmatrix} B'_1 & . & . & B'_{NP} \end{bmatrix} \begin{bmatrix} X_1 \\ . \\ . \\ X_{NP} \end{bmatrix}
$$

$$
+ \begin{bmatrix} B'_1 & . & . & B'_{NP} \end{bmatrix} \begin{bmatrix} z^{-1} \\ . \\ . \\ z^{-NP} \end{bmatrix} \begin{bmatrix} A_1 & . & . & A_{NP} \end{bmatrix} \begin{bmatrix} X_1 \\ . \\ . \\ X_{NP} \end{bmatrix}
$$

$$(6.109)$$

Cancelling out $\mathbf{B}'^T = \begin{bmatrix} B'_1 .. B'_{NP} \end{bmatrix}$ we get

$$
V^i \begin{bmatrix} z^{-1} \\ . \\ . \\ z^{-NP} \end{bmatrix} = \begin{bmatrix} X_1 \\ . \\ . \\ X_{NP} \end{bmatrix} + \begin{bmatrix} z^{-1} \\ . \\ . \\ z^{-NP} \end{bmatrix} \begin{bmatrix} A_1 & . & . & A_{NP} \end{bmatrix} \begin{bmatrix} X_1 \\ . \\ . \\ X_{NP} \end{bmatrix} \qquad (6.110)
$$

The first row of (6.110) gives state variable X_1 in terms of previous values

$$
X_1 = z^{-1} V^i - z^{-1} \sum_{i=1}^{NP} A_i X_i \qquad (6.111)
$$

From the second row of (6.106) we get

$$
X_2 = z^{-1} \left(z^{-1} V^i - z^{-1} \sum_{i=1}^{NP} A_i X_i \right) = z^{-1} X_1 \qquad (6.112)
$$

and so on for the remaining state variables. Equation (6.108) is the output equation and in compact form is expressed as

$$
V^r = B_0 V^i + \mathbf{B}'^T X \qquad (6.113)
$$

Equations (6.111), (6.112), etc., may be put in compact form to form the state-space equation.

$$
\mathbf{X} = z^{-1} \mathbf{A}' \mathbf{X} + z^{-1} \mathbf{1}' V^i \qquad (6.114)
$$

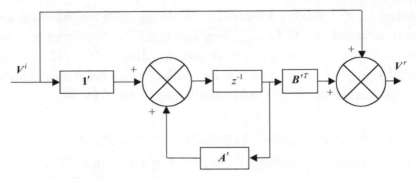

FIGURE 6.12: General flow diagram showing the implementation of frequency-dependent scattering in the time-domain.

where

$$A' = \begin{bmatrix} -A_1 & -A_2 & . & . & -A_{\text{NP}-1} & -A_{\text{NP}} \\ 1 & & & & & \\ & 1 & & & & \\ & & . & & & \\ & & & . & & \\ & & & & 1 & \end{bmatrix} \quad 1' = \begin{bmatrix} 1 \\ 0 \\ . \\ . \\ . \\ 0 \end{bmatrix} \quad (6.115)$$

Equation (6.113) and (6.114) describe a procedure in the time-domain for implementing the frequency-dependent scattering. It is summarized in the graph shown in Fig. 6.12. A scheme like this may be used to implement each scattering event.

It remains to comment on the way the general scattering function in (6.101) may be obtained. The general procedure is to start from the samples of the scattering coefficients in the frequency-domain. This information may be obtained analytically, from measurements or from specific detailed numerical studies. The frequency-domain Prony method is then employed to obtain a least-square approximation in terms of exponential functions

$$f(t) = \sum_{i=0}^{\text{NP}-1} C_i e^{s_{pi} t} \quad (6.116)$$

The number of terms NP in this expansion depends on the complexity of the response, the accuracy required, and numerical stability issues. The Laplace transform of (6.116) is

$$F(s) = \sum_{i=0}^{\text{NP}-1} \frac{C_i}{s - s_{pi}} \quad (6.117)$$

Equation (6.117) may then be put in the form of (6.102) and the task then is to obtain the a and b coefficients in the Padé approximation from knowledge of a number of data samples in the frequency-domain. The detailed procedure for calculating these coefficients from frequency-domain data is described in [50, 52, 53]. Very significant computational cost reductions are obtained compared to a case where a fine mesh is used to describe in detail each perforation. In a similar fashion thin panels, other than metallic, made out of layers of carbon fiber composites or other materials may be modeled. Further details may be found in [54].

6.7   DEALING WITH FINE OBJECTS IN 3D TLM

In section 4.6 we addressed the problem of embedding thin wires in a 2D TLM mesh when the radius of the wire is smaller than the spatial resolution of the mesh. We explained the practical need for this and ways in which it could be done. It will come as no surprise that in 3D it is even more pressing to adopt similar multiscale techniques for modeling fine features. The approach is similar to that adopted for 2D calculations. We express local solutions in terms of modes but since we now have more degrees of freedom (12 in a standard SCN) we can accommodate more modes. We then seek to calculate the impedance seen by each mode. We can then proceed in one of the two ways. One approach is to develop a circuit such that it presents to each mode the correct impedance. The total incident voltages at the ports of the node are then automatically decomposed into incident mode voltages which after encountering the appropriate impedance return as reflected modal components and automatically recombine to produce the total reflected voltage pulses which connect to the rest of the mesh in a typical TLM fashion. Alternatively, the decomposition into modes and the subsequent recomposition may be done as a signal-processing task without resort to a circuit. Details of how this is done in 3D for conducting wires placed at the center of the node may be found in [55]. The classical approach for a single wire and a multiconductor system is explained in [56, 57].

The case of an arbitrary inclined wire has also been dealt with in [58]. It can also be applied to describe embedded features other than wires such as conducting or dielectric spherical features [59].

If an analytical expression for the modes of an embedded feature is not available then these can be obtained by numerical means [60].

Finally, unstructured meshes may also be developed for 3D TLM generalizing the techniques described in Chapter 5 for 2D TLM. The basic element here is the tetrahedron and as for the 2D case a modal expansion allows the development of an appropriate circuit to implement scattering for this very general node. A detailed description is not within the scope of the present text. Details may be found in [61].

6.8   THEORETICAL FOUNDATIONS OF TLM AND ITS RELATION TO OTHER METHODS

We have discussed the basics of TLM and how it is being developed to deal with challenging technical problems in an efficient and accurate manner. Examples were multiscale techniques (wires, thin panels, apertures) and advanced materials modeling. Whenever possible I have chosen to give simple intuitive explanations to illustrate the modeling concepts. In this way I hope that I have been able to convey the essence of TLM and to give readers a rapid introduction to the subject and the ability to develop their own code. At the same time I have made several in-depth studies of more complex issues so that the reader gets a true sense of the research frontiers and the prospects for TLM. It is inevitable that I have omitted several topics which although important either would not fit the pattern of the presentation or could not be presented in a meaningful way within the constraints of a text of this size. In this final section I would like to make a brief reference to these topics so that the interested reader can consult the relevant literature.

The original derivation of the 3D TLM scheme was done by analogy to a network of transmission lines. This can be seen in the work of Kron [62] and also in [4, 32]. However, the reader may wish to see a more systematic derivation of the TLM scheme for methodological reasons and also as it makes it easier to see the relationship of TLM with other computational electromagnetic methods and to study its dispersion properties. Several workers have addressed this issue in different ways. The finite-difference modeling of Maxwell's equation with the SCN TLM node is described in [63] and an approach based on finite integration is developed in [64]. The finite integration scheme [65, 66] proceeds by defining the vector of the 12 incident voltages at time k and node (l, m, n) $_k\boldsymbol{a}_{l,m,n}$ and the reflected voltage vector $_k\boldsymbol{b}_{l,m,n}$ where we have used the notation of [6]. The electric and magnetic field vectors are then obtained from

$$
\begin{aligned}
_k\boldsymbol{E}_{l.m.n} &= \sqrt{Z_{TL}}\,(_k\boldsymbol{a}_{l,m,n} + {}_k\boldsymbol{b}_{l,m,n}) \\
_k\boldsymbol{H}_{l.m.n} &= \sqrt{Z_{TL}}\,(_k\boldsymbol{a}_{l,m,n} - {}_k\boldsymbol{b}_{l,m,n})
\end{aligned}
\tag{6.118}
$$

We then apply Ampere's law and Faraday's law, where the Crank–Nicholson scheme is used for time disretization, on surfaces through the node center to obtain 12 independent equations in total, relating the reflected voltage pulses to the incident pulses. After some algebra we can obtain the TLM scattering matrix already derived in the previous chapter.

Yet another way to proceed is to derive the TLM scheme from the method of moments [67, 68]. Here the electric and magnetic fields are expanded in terms of known spatial and time basis functions with unknown coefficients. These expansions are then substituted into Maxwell's equations that are sampled using delta test functions to recover the original TLM scattering equations.

I also note the work to derive TLM from modal considerations which was described in some detail in previous chapters. This gives additional insight into the modeling process.

Substantial work has been done to relate and compare TLM to FDTD both for methodological purposes and to establish the dispersion characteristics of each technique [69–71]. Dispersion properties may be established by numerical studies but closed form expressions for a number of cases are also available [72–74].

Of considerable interest in computational electromagnetics is the issue of absorbing boundary conditions (ABC). As already mentioned the matching BC in TLM is a good and computationally inexpensive option. However, in very demanding applications a better ABC may be necessary. In problems with simple open boundaries the concept of John's matrix described in [75] is a useful one but can become computationally expensive in general problems. Approaches based on the "perfectly matched layer" PML that are widely investigated in connection with other numerical methods may also be applied in TLM. Further details may be found in [76]. As a result of further work on TLM, alternatives to the standard SCN have been proposed. Of particular note are the Alternating TLM (ATLM) and the Alternate-Direction-Implicit TLM (ADI-TLM). Work by researchers at the Technical University of Munich has offered further insight into the processes of TLM schemes [68, 77]. They have shown that by a suitable rotation two independent six-ports, which are interleaved but not coupled, may be obtained. There are thus two "parity" states and we can employ one and obtain the other by interpolation (the ATLM scheme). This in theory is attractive as it eliminates spurious solutions and can halve the computational costs. However, the different parity states couple at boundaries, discontinuities, and stubs and this means that this scheme has not found favor in practice. Spurious solutions in the standard SCN may be avoided by smoother excitation (i.e., not exciting at a single node but instead spreading excitation over several nodes).

The TLM schemes described so far are explicit schemes—the state of a TLM node at a given time step depends only on the state of neighboring nodes at the previous time step. This provides for a simple algorithm but the time step is set by physical considerations and hence it is generally small. This is computationally expensive for some applications. The ADI-TLM or Split-Step TLM (SS-TLM) overcomes this by splitting sampling in time into two steps. The resulting algorithm is implicit but we can employ a larger time step [78].

I also mention techniques for generating hybrid models (e.g., TLM and IE methods) which may be employed to increase overall model efficiency [79].

Appendix 1

MAXWELL'S EQUATIONS

Ampere's law,

$$\nabla \times \boldsymbol{H} = \boldsymbol{J} + \frac{\partial \boldsymbol{D}}{\partial t}$$

Faraday's law,

$$\nabla \times \boldsymbol{E} = -\frac{\partial \boldsymbol{B}}{\partial t}$$

Gauss's law,

$$\nabla \cdot \boldsymbol{D} = \rho$$

Magnetic flux conservation,

$$\nabla \cdot \boldsymbol{B} = 0$$

Ampere's law and Faraday's law in Cartesian coordinates:

$$\frac{\partial H_z}{\partial y} - \frac{\partial H_y}{\partial z} = J_x + \frac{\partial D_x}{\partial t}$$

$$\frac{\partial H_x}{\partial z} - \frac{\partial H_z}{\partial x} = J_y + \frac{\partial D_y}{\partial t}$$

$$\frac{\partial H_y}{\partial x} - \frac{\partial H_x}{\partial y} = J_z + \frac{\partial D_z}{\partial t}$$

$$\frac{\partial E_z}{\partial y} - \frac{\partial E_y}{\partial z} = -\frac{\partial B_x}{\partial t}$$

$$\frac{\partial E_x}{\partial z} - \frac{\partial E_z}{\partial x} = -\frac{\partial B_y}{\partial t}$$

$$\frac{\partial E_y}{\partial x} - \frac{\partial E_x}{\partial y} = -\frac{\partial B_z}{\partial t}$$

Appendix 2

SMALL ARGUMENT EXPANSIONS FOR BESSEL FUNCTIONS

The following expansions are for $z \ll 1$:

$$J_0(z) \simeq 1 - \frac{z^2}{4}$$

$$J_1(z) \simeq \frac{z}{2}$$

$$J_2(z) \simeq \frac{z^2}{8}$$

$$J_n(z) \simeq \frac{\left(\frac{z}{2}\right)^n}{n!}, \quad n \neq 0$$

$$\frac{dJ_0(z)}{dz} \simeq -\frac{z}{2}$$

$$\frac{dJ_1(z)}{dz} \simeq \frac{1}{2}$$

$$\frac{dJ_2(z)}{dz} \simeq \frac{z}{4}$$

$$\frac{dJ_n(z)}{dz} \simeq \frac{z^{n-1}}{2^n (n-1)!}$$

$$N_0(z) \simeq \frac{2}{\pi} \left[\ln\left(\frac{z}{2}\right) + \gamma \right], \quad \gamma = 0.577$$

$$N_1(z) \simeq -\frac{2}{\pi z}$$

$$N_2(z) \simeq -\frac{4}{\pi z^2}$$

$$N_n(z) \simeq -\frac{(n-1)!}{\pi} \left(\frac{2}{z}\right)^n, \quad n \neq 1$$

$$\frac{dN_0(z)}{dz} \simeq \frac{2}{\pi z}$$

$$\frac{d N_1(z)}{dz} \simeq \frac{2}{\pi z^2}$$

$$\frac{d N_2(z)}{dz} \simeq \frac{8}{\pi z^3}$$

$$\frac{d N_n(z)}{dz} \simeq \frac{n! 2^n}{\pi} \frac{1}{z^{n+1}}$$

References

[1]   E. K. Miller, "A selective survey of computational electromagnetics," *IEEE Trans. Antennas Propag.*, vol. 26, pp. 1281–1305, 1988.doi:10.1109/8.8607

[2]   K. S. Kunz and R. J. Luebbers, *The Finite Difference Time Domain Method for Electromagnetics*, CRC Press, Boca Raton, 2000.

[3]   A. Taflove, *Computational Electrodynamics: The Finite Difference Time-Domain Method*, Artech House, Boston, 1995

[4]   C. Christopoulos, *The Transmission-Line Modeling Method: TLM*, IEEE Press, New York, 1995.

[5]   R. F. Harrington, *Field Computation by Moment Methods*, IEEE Press, New York, 1968.

[6]   P. Russer, *Electromagnetics, Microwave Circuit and Antenna Design for Communications Engineering*, Artech House, Boston, 2003.

[7]   A. Hirose and K. E. Lonngren, *Introduction to Wave Phenomena*, Wiley, New York, 1985.

[8]   S. Ramo, J. R. Whinnery, and T. van Duzer, *Fields and Waves in Communication Electronics*, Wiley, New York, 1984.

[9]   S. Y. Hui and C. Christopoulos, "Discrete transform technique for solving non-linear circuits and equations," *Proc. IEE-A*, vol. 137, pp. 379–384, 1992.

[10] C. Huygens, *Traite de la Lumiere*, Paris, Leiden, 1690.

[11] P. B. Johns, "A new mathematical model to describe the physics of propagation," *Radio and Electronic Engineer*, vol. 44, pp. 657–666, 1974.

[12] —— "The solution of inhomogeneous waveguide problems using a transmission-line matrix," *IEEE Trans. Microw. Theory Tech.*, vol. 22, pp. 284–288, 1974.

[13] W. J. R. Hoefer and P. P. M. So, *The Electromagnetic Wave Simulator*, Wiley, New York, 1991.

[14] J. A. Morente, J. A. Porti, and M. Khalladi, "Absorbing boundary conditions for the TLM method," *IEEE Trans. Microw. Theory Tech.*, vol. 40, pp. 2095–2099, 1992. doi:10.1109/22.168768

[15] C. A. Balanis, *Advanced Engineering Electromagnetics*, Wiley, New York, 1989.

[16] Y. K. Choong, *Advanced Modal Expansion Techniques for the TLM Method*, Dr. Phil. Thesis, University of Nottingham, Oct. 2003.

[17] C. J. Smartt, private communication, 2005

[18] L. R. Rabiner and B. Gold, *Theory and Application of Digital Signal Processing*, Prentice-Hall, New Jersey, 1975.

[19] C. Christopoulos, "Multi-scale modelling in time-domain electromagnetics," *Int. J. Electron. Commun. (AEÜ)*, vol. 57, no 2, pp. 100–110, 2003.

[20] R. Holland and L Simpson, "Finite-difference analysis of EMP coupling in thin struts and wires," *IEEE Trans. Electromagn. Compat.*, vol. 23, pp. 88–97, 1981.

[21] Y. K. Choong, P. Sewell, and C. Christopoulos, "Accurate modelling of an arbitrarily placed thin wire in a coarse mesh," *IEE Proc.—Sci. Meas. Technol.*, vol. 149, no 5, pp. 250–153, 2002.doi:10.1049/ip-smt:20020590

[22] K. Biwojno, P. Sewell, Y. Liu, and C. Christopoulos, "Embedding multiple wires in a single TLM node," in *Proc. EUROEM*, Magdeburg, 2004, pp. 173–174.

[23] C. Christopoulos, P. Sewell, K. Biwojno, Y. Liu, and C. J. Smartt, "Hierarchical models of complex systems in the time-domain," in *Proc. XXVIII General Assembly of the Union of Radio Science (URSI)*, Delhi, 2005, paper EO1.4, 4 pages.

[24] K. Biwojno, P. Sewell, Y. Liu, and C. Christopoulos, "Electromagnetic modelling of fine features in photonic applications," in *Proc. Optical Waveguide Theory and Numerical Modelling*, Grenoble, 2005, paper SA3.4.

[25] D. Al-Muktar and J. Sitch, "Transmission-line matrix method with irregularly graded space," *Proc IEE, Part H*, vol. 128, pp. 299–305, 1981.

[26] R. Allan and R. J. Clark, "Application of the TLM method to the cold modelling of magnetrons," *Int. J. Numer. Model.*, vol. 1, pp. 221-238, 1988.doi:10.1002/jnm.1660010406

[27] A. C. Cangellaris and D. B. Wright, "Analysis of the numerical error caused by the stair-stepped approximation of a conducting boundary in FDTD simulations of electromagnetic phenomena," *IEEE Trans. Antennas Propag.*, vol. 39, pp. 1518–1525, 1991. doi:10.1109/8.97384

[28] P. Sewell, J. Wykes, T. M. Benson, C. Christopoulos, D. W. P. Thomas, and A. Vukovic, *IEEE Trans. Microw. Theory Tech.*, vol. 52, no 5, pp. 1490–1497, 2004. doi:10.1109/TMTT.2004.827027

[29] P. M. Chirlian, *Basic Network Theory*, McGraw-Hill, 1969.

[30] J. R. Schewchuk, *Triangle: A two-dimensional Quality Mesh Generator and Delaunay Triangulator*, www-2.cs.cmu.edu/~quake/triangle.html

[31] P. Sewell, J. Wykes, A. Vukovic, D. W. P. Thomas, T. M. Benson and C. Christopoulos, "Multi-grid interface in computational electromagnetics," *Electron. Lett.*, vol. 40, no 3, pp. 162–163, 2004.doi:10.1049/el:20040107

[32] P. B. Johns, "A symmetrical condensed node for the TLM method," *IEEE Trans. Microw. Theory Tech.*, vol. 35, pp. 370–377, 1987.doi:10.1109/TMTT.1987.1133658

[33] J. L. Herring, *Developments in the Transmission-Line Modelling Method for Electromagnetic Compatibility Studies*, Dr. Phil. Thesis, University of Nottingham, May 1993.

[34] V. Trenkic, *The Development and Characterization of Advanced Nodes for the TLM Method*, Dr. Phil. Thesis, University of Nottingham, Nov. 1995.

[35] V. Trenkic, C. Christopoulos, and T. M. Benson, "Simple and elegant formulation of scattering in TLM nodes," *Eletron. Lett.*, vol. 29, no. 18, pp. 1651–1652, 1993.

[36] F. J. German, "Infinitesimally adjustable boundaries in symmetrical condensed node TLM simulations," in *9th Annual Rev. of Progress in Applied Comp. Electromagnetics*, NPS Monterey, pp. 483–490, 1993.

[37] R. A.. Scaramuzza and A. J. Lowery, "Hybrid symmetrical condensed node for the TLM method," *Electron. Lett.*, vol. 26, pp. 1947–1949, 1990.

[38] V. Trenkic, C. Christopoulos, and T. M. Benson, "On the time step in hybrid symmetrical condensed TLM nodes," *IEEE Trans. Microw. Theory Tech.*, vol. 43, no. 9, pp. 2172–2174, 1995.doi:10.1109/22.414558

[39] P. Berrini and K. Wu, "A pair of hybrid symmetrical condensed TLM nodes," *IEEE Microw. Guided Wave Lett.*, vol. 4, no. 7, pp. 244–246, 1994.doi:10.1109/75.298254

[40] V. Trenkic, C. Christopoulos, and T. M. Benson, "Advanced node formulations in TLM-The matched symmetrical condensed node (MSCN)," in *12th Annual Rev. of Progress in Appl. Comp. Electromagnetics*, NPS, Monterey, 1996, pp. 246–253.

[41] —— "Advanced node formulations in TLM- The adaptable symmetrical condensed node (ASCN)," *IEEE Trans. Microw. Theory Tech.*, vol. 44, no. 12, pt. II, pp. 2473–2478, 1996.doi:10.1109/22.554580

[42] D. P. Johns, A. J. Wlodarczyk, A. Mallik, and C Christopoulos, "New TLM technique for steady-state field solutions in three-dimensions," *Electron. Lett.*, vol. 28, pp. 1692–1694, 1992.

[43] H. Jin and R. Vahldieck, "The frequency-domain transmission-line matrix method—a new concept," *IEEE Trans. Microw. Theory Tech.*, vol. 40, pp. 2207–2218, 1992. doi:10.1109/22.179882

[44] J. F. Dawson, "Representing ferrite tiles as frequency dependent boundaries in TLM," *Electron. Lett.*, vol. 29, no. 9, pp. 791–792, 1993.

[45] L. de Menezes and W. J. R. Hoefer, "Modeling of general constitutive relationships using SCN TLM," *IEEE Trans. Microw. Theory Tech.*, vol. 44, pp. 854–861, 1996. doi:10.1109/22.506444

[46] I. Barba, J Represa, M. Fujii, and W. J. R. Hoefer, "Multiresolution model of electromagnetic wave propagation in dispersive materials," *IEEE MTT-S, Int. Microw. Symp. Dig.*, Anaheim, pp. 1471–1474, 1999.

[47] J. D. Paul, C. Christopoulos, and D. W. P. Thomas, "Generalised material models in TLM—Part 1: Materials with frequency dependent properties," *IEEE Trans. Antennas Propag.*, vol. 47, pp. 1528–1534, 1999.doi:10.1109/8.805895

[48] —— "Generalised material models in TLM—Part 2: Materials with anisotropic properties properties," *IEEE Trans. Antennas Propag.*, vol. 47, pp. 1535–1542, 1999. doi:10.1109/8.805896

[49] —— "Generalised material models in TLM—Part 1:Materials with non-linear properties," *IEEE Trans. Antennas Propag.*, vol. 50, pp. 997–1004, 2002. doi:10.1109/TAP.2002.800733

[50] J. D. Paul, *Modelling of General Electromagnetic Material Properties in TLM*, Dr. Phil. Thesis, University of Nottingham, Oct. 1998.

[51] C. Christopoulos, *Principles and Techniques of Electromagnetic Compatibility*, CRC Press, Boca Raton, 1995.

[52] J. N. Brittingham, E. K. Miller, and J. L. Willows, "Pole extraction from real frequency information," *Proc. IEEE*, vol. 68, no. 2, pp. 263–273, 1980.

[53] J. D. Paul, C. Christopoulos, and D. W. P. Thomas, "Equivalent circuit models for the time-domain simulation of ferrite electromagnetic wave absorbers," *in Proc. 13th Int. Zurich EMC Symp.*, Zurich, 1999, pp. 345–350.

[54] J. D. Paul, V. Podlozny, and C. Christopoulos, "The use of digital filtering techniques for the simulation of fine features in EMC problems solved in the time-domain," *IEEE Trans. Electromagn. Compat.*, vol. 45, no. 2, pp. 238–244, 2003. doi:10.1109/TEMC.2003.810810

[55] P. Sewell, Y. K. Choong, and C. Christopoulos, "An accurate thin-wire model for 3D TLM simulations," *IEEE Trans. Electromagn. Compat.*, vol. 45, no. 2, pp. 207–217, 2003. doi:10.1109/TEMC.2003.810812

[56] J. D. Paul, C. Christopoulos, D. W. P. Thomas, and X. Liu, " Time-domain modelling of electromagnetic wave interaction with thin-wires using TLM," *IEEE Trans. Electromagn. Compat.*, vol. 47, no. 3, pp. 447–455, 2005.doi:10.1109/TEMC.2005.852217

[57] A. J. Wlodarczyk, V. Trenkic, R. A. Scaramuzza, and C. Christopoulos, "A fully integrated multi-conductor model for TLM," *IEEE Trans. Microw. Theory Tech.*, vol. 46, no. 2, pp. 2431–2437, 1998.doi:10.1109/22.739231

[58] Y. Liu, P. Sewell, K. Biwojno, and C Christopoulos, " A generalised node for embedding sub-wavelength objects into 3D TLM," *IEEE Trans. Electromagn. Compat.*, to appear.

[59] K. Biwojno, P. Sewell, Y. Liu, and C. Christopoulos, "Embedding multiple wires within a single TLM node," in *Ultra-Wideband Short-Pulse Electromagnetics (UWB SP7)*, to appear.

[60] K. Biwojno, C. J. Smartt, P. Sewell, Y. Liu, and C. Christopoulos, "General treatment of TLM node with embedded structures," *Int. J. Numer. Model.*, to appear., 2006.

[61] P. Sewell, T. M. Benson, C. Christopoulos, D. W. P. Thomas, A. Vukovic, and J Wykes, "Transmission-Lime Modelling (TLM) based upon tetrahedral meshes," *IEEE Trans. Microw. Theory Tech.*, vol. 53, no. 6, pp. 1919–1928, 2005. doi:10.1109/TMTT.2005.848078

[62] G. Kron, "Equivalent circtuit of the field equations of Maxwell," *Proc. Inst. Radio Eng.*, vol. 32, pp. 284–288, 1944.

[63] S. Hein, "Consistent finite-difference modelling of Maxwell's equations with lossy symmetrical condensed TLM node," *Int. J. Numer. Model.*, vol. 6, pp. 207–220, 1993. doi:10.1002/jnm.1660060305

[64] M. Aidam and P. Russer, "Derivation of the TLM method by finite integration," *Int. J. Electron. Commun. (AEÜ)*, vol. 51, pp. 35–39, 1997.

[65] N. Pena and M. Ney, " A general and complete two-dimensional TLM hybrid node formulation based on Maxwell's integral equations," in *Proc. 12th Annual Review of Progress in Applied Comp. Electromagnetics*, NPS, Monterey, 1996, pp. 254–261.

[66] —— "A general formulation of a three-dimensional TLM condensed node with the modelling of electric and magnetic losses and current sources," in *Proc. 12th Annual Review of Progress in Applied Comp. Electromagnetics*, 1996, pp. 262–269.

[67] M. Krumpholz and P. Russer, "A field theoretical derivation of TLM," *IEEE Trans. Microw. Theory Tech.*, vol. 42, pp. 1660–1668, 1994.doi:10.1109/22.310559

[68] P. Russer and C. Christopoulos, *Part IV in Applied Computational Electromagnetics, NATO ASI Series F*, vol. 171, N. K. Uzunoglu, K. S. Nikita, and D. I. Kaklamani, Eds., Springer, Berlin, 1998.

[69] Z. Chen, M. Ney, and W. J. R. Hoefer, " A new finite-difference time-domain formulation and its equivalence to the TLM SCN," *IEEE Trans. Microw. Theory Tech.*, vol. 39, pp. 2160–2169, 1991.doi:10.1109/22.106559

[70] F. J. German, J. A. Svigelj, and R. Mittra, "A numerical comparison of dispersion in irregularly graded TLM and FDTD meshes," in *Proc. 12th Annual Review of Progress in Applied Comp. Electromagnetics*, 1996, pp. 270–278.

[71] N. R. S. Simons, R. Siushansian, J. Lo Vetri, and M. Cuhaci, "Comparison of TLM and FDTD methods for a problem containing a sharp metallic edge," *IEEE Trans. Microw. Theory Tech.*, vol. 47, no. 10, pp. 2042–2045, 1999.doi:10.1109/22.795084

[72] J. S. Nielsen and W. J. R. Hoefer, " A complete dispersion analysis of the condensed node TLM mesh," *IEEE Trans. Magn.*, vol. 27, pp. 3982–3985, 1991.doi:10.1109/20.104974

[73] M. Krumpholz and P. Russer, "On the dispersion in TLM and FDTD," *IEEE Trans. on Microw. Theory Tech.*, vol. 42, pp. 1275–1279, 1994.doi:10.1109/22.299768

[74]   V. Trenkic, C. Christopoulos, and T. M. Benson, "Analytical expansion of the dispersion relation for TLM condensed nodes," *IEEE Trans. on Microw. Theory Tech.*, vol. 44, pp. 2223–2230, 1996.doi:10.1109/22.556450

[75]   W. J. R. Hoefer, " The discrete time-domain Green's function or John's matrix- a new powerful concept in TLM," *Int. J. Numer. Modeling*, vol. 2, pp. 215–225, 1989. doi:10.1002/jnm.1660020405

[76]   S. Le Maguer and M. M. Ney, "Extended PML-TLM: An efficient approach for full-wave analysis of open structures," *Int. J. Numer. Modeling*, vol. 14, pp. 129–144, 2001. doi:10.1002/jnm.402

[77]   P. Russer and B. Bader, "The alternating transmission-line matrix (ATLM) scheme," in *IEEE Int. Microw. Symp. Dig.*, Orlando, 1995, pp. 19–22.

[78]   S. Le Maguer and M. M. Ney, "Split step TLM (SS-TLM), for efficient electromagnetic simulation of small heterogeneous apertures," in *Proc. Int. Zurich EMC Symp.*, Zurich, 2003, pp. 275–278.

[79]   S. Lindenmeir, L. Pierantoni, and P. Russer, "Hybrid space discretizing integral equation methods for numerical modelling of transient interference," *IEEE Trans. Electromagn. Compat.*, vol. 41, pp. 425–430, 1999.doi:10.1109/15.809843

Author Biography

Christos Christopoulos was born in Patras, Greece, on September 17, 1946. He received the Diploma in Electrical and Mechanical Engineering from the National Technical University of Athens in 1969 and the M.Sc. and D.Phil. degrees from the University of Sussex in 1979 and 1974, respectively.

In 1974, he joined the Arc Research Project of the University of Liverpool and spent 2 years working on vacuum arcs and breakdown while on attachments at the UKAEA Culham Laboratory. In 1976 he joined the University of Durham as a Senior Demonstrator in Electrical Engineering Science. In October 1978 he joined the Department of Electrical and Electronic Engineering, University of Nottingham, where he is now Professor of Electrical Engineering and Director of the George Green Institute for Electromagnetics Research (GGIEMR).

His research interests are in computational electromagnetics, electromagnetic compatibility, signal integrity, protection and simulation of power networks, and electrical discharges and plasmas. He is the author of over 300 research publications and five books. He has received the Electronics Letters and the Snell Premiums from the IEE and best paper awards of several conferences. He is a member of the IEE, IoP and an IEEE Fellow. Formerly, he was Executive Team Chairman of the IEE Professional Network in EMC, Member of the CIGRE Working Group 36.04 on EMC, and Associate Editor of the IEEE EMC Transactions. He is Vice-Chairman of URSI Commission E "Noise and Interference" and Associate Editor of the URSI Radio Bulletin.

Printed in the United States
by Baker & Taylor Publisher Services